Synthesis Lectures on Mobile and Pervasive Computing

Editor
Mahadev Satyanarayanan, *Carnegie Mellon University*

Synthesis Lectures on Mobile and Pervasive Computing is edited by Mahadev Satyanarayanan of Carnegie Mellon University. Mobile computing and pervasive computing represent major evolutionary steps in distributed systems, a line of research and development that dates back to the mid-1970s. Although many basic principles of distributed system design continue to apply, four key constraints of mobility have forced the development of specialized techniques. These include: unpredictable variation in network quality, lowered trust and robustness of mobile elements, limitations on local resources imposed by weight and size constraints, and concern for battery power consumption. Beyond mobile computing lies pervasive (or ubiquitous) computing, whose essence is the creation of environments saturated with computing and communication, yet gracefully integrated with human users. A rich collection of topics lies at the intersections of mobile and pervasive computing with many other areas of computer science.

Intelligent Notification Systems
Abhinav Mehrotra and Mirco Musolesi
2020

Privacy in Mobile and Pervasive Computing
Marc Langheinrich and Florian Schaub
2018

Mobile User Research: A Practical Guide
Sunny Consolvo, Frank R. Bentley, Eric B. Hekler, and Sayali S. Phatak
2017

Pervasive Displays: Understanding the Future of Digital Signage
Nigel Davies, Sarah Clinch, and Florian Alt
2014

Cyber Foraging: Bridging Mobile and Cloud Computing
Jason Flinn
2012

Intelligent Notification Systems

Intelligent Notification Systems

Abhinav Mehrotra and Mirco Musolesi

ISBN: 978-3-031-01359-1   paperback
ISBN: 978-3-031-02487-0   ebook
ISBN: 978-3-031-00316-5   hardcover

DOI: 10.1007/978-3-031-02487-0

A Publication in the Springer series
SYNTHESIS LECTURES ON MOBILE AND PERVASIVE COMPUTING

Lecture #14
Series Editor: Mahadev Satyanarayanan, *Carnegie Mellon University*
Series ISSN
Print 1933-9011   Electronic 1933-902X

Intelligent Notification Systems

Abhinav Mehrotra
Samsung AI Center, Cambridge, UK

Mirco Musolesi
University College London, UK and
University of Bologna, Italy

SYNTHESIS LECTURES ON MOBILE AND PERVASIVE COMPUTING #14

ABSTRACT

Notifications provide a unique mechanism for increasing the effectiveness of real-time information delivery systems. However, notifications that demand users' attention at inopportune moments are more likely to have adverse effects and might become a cause of potential disruption rather than proving beneficial to users. In order to address these challenges a variety of intelligent notification mechanisms based on monitoring and learning users' behavior have been proposed. The goal of such mechanisms is maximizing users' receptivity to the delivered information by automatically inferring the right time and the right context for sending a certain type of information. This book presents an overview of the current state of the art in the area of intelligent notification mechanisms that rely on the awareness of users' context and preferences. We first present a survey of studies focusing on understanding and modeling users' interruptibility and receptivity to notifications from desktops and mobile devices. Then, we discuss the existing challenges and opportunities in developing mechanisms for intelligent notification systems in a variety of application scenarios.

KEYWORDS

notification systems, interruptibility, context-aware computing, anticipatory computing, intelligent mobile systems, intelligent user interfaces

Contents

Preface

Intelligent notification systems are the key building blocks of modern computing systems. Indeed, from mobile phones to desktop computers and from personal assistants to wearable devices, notifications are becoming a key part of our interactions with interactive systems. With the continuous integration of machine learning and artificial intelligence algorithms in digital systems, the underlying components for deciding what, when, and where to send notifications are becoming more and more intelligent. This book provides an in-depth introduction to the state of the art of this fascinating field. It is aimed at both researchers and practitioners working in this area or interested in exploring this highly cross-disciplinary topic.

The book only assumes a very basic knowledge of human-computer interaction and ubiquitous computing concepts. In general, several definitions are provided and extensive bibliography can be found at the end of book. The structure and contents of this book can be summarized as follows.

- In Chapter 1 (*Introduction*), we introduce the scope of the book and its aims, outlining the key issues in designing intelligent notification systems.

- In Chapter 2 (*Understanding Users' Interaction with Notifications*), we frame the problem by examining possible definitions of *interruptions* based on past and current theories, give an overview of their types and discuss human response to such interruptions.

- In Chapter 3 (*Costs Associated to Interruptions*), we present the findings of various studies that examined the cost associated with the arrival of interruptions at inopportune moments. More specifically, we discuss the detrimental effects of interruptions on users' memory, emotional and affective states, and ongoing task execution.

- In Chapter 4 (*Personalization of Interruptibility Strategies*), we provide an overview of the findings of studies concerning personalization of interruptibility strategies. We focus in particular on the effectiveness of personalization strategies with respect to time allocation in a multitasking environment. Moreover, we report the findings of studies demonstrating that the performance in carrying out an interrupted task is affected by users' anxiety and arousal levels.

- In Chapter 5 (*Design Principles for Interruptibility Management Systems*), we analyze the design of interruption management systems and discuss the difference between users' attentiveness and receptivity to interruptions.

- In Chapter 6 (*Interruptibility Management Systems for Desktop Environments*), we first introduce the characteristics of interruptions in desktop environment as they arrive in a constant context (i.e., desktops are not taken around with users). We then discuss the studies in the area of interruptibility management for desktop environment that were conducted by using a Wizard of Oz approach, by exploiting task phases and through on-the-fly inference of interruptibility.

- In Chapter 7 (*Interruptibility Management Systems for Mobile Environments*), we first discuss how the advent of mobile phones has provided opportunities for users to connect to different information channels and receive updates in real time about a variety of events. We then discuss the studies in the area of interruptibility management for mobile environment based on the analysis of current activity, the transition between activities, contextual data, and those concerning filtering of irrelevant information.

- In Chapter 8 (*Open Challenges and Outlook*), we first provide an overview of the limitations of existing studies that focus on understanding and learning users' behavior in terms of interactions with notifications. We then summarize the key elements of this lecture outlining the open challenges in the field.

We hope that this book will stimulate further developments in this field. In the end, the acceptance of Artificial Intelligence technologies also depends on the design of efficient and effective intelligent solutions for the delivery of information to the end-users. Notifications are an essential class of interaction modality for intelligent technologies. For this reason, we believe a strong foundation in this area is important for everyone involved in the design of such systems.

Abhinav Mehrotra and Mirco Musolesi
December 2019

Acknowledgments

The authors would like to acknowledge Rami Bahsoon, Russell Beale, Rowanne Fleck, Robert Hendley, Per Ola Kristensson, and Veljko Pejovic for the useful comments and discussions about the content of this book.

This work was supported through the EPSRC grants EP/P016278/1 and EP/L018829/2 at UCL and also by The Alan Turing Institute under the EPSRC grant EP/N510129/1.

Abhinav Mehrotra and Mirco Musolesi
December 2019

CHAPTER 1

Introduction

1.1   MOBILE NOTIFICATIONS

Mobile phones represent an essential element of our lives by assisting us in several day-to-day activities. Since they are always connected to the Internet, mobile phones represent a unique platform for receiving or fetching information anytime and anywhere. This is leveraged by numerous mobile applications, such as email and instant messenger clients, VoIP (Voice over Internet Protocol) services, and social network platforms to provide their core functionalities.

The key to success of such applications, which essentially provide access to a variety of information channels, is to ensure real-time awareness of users about the delivered information. In order to ensure this, mobile operating systems facilitate the use of *notifications* (as shown in Figure 1.1) that steer users' attention toward the delivered information through audio, visual, and haptic signals. This is indeed in contrast with the traditional paradigm of *pull-based* information retrieval and delivery in which the user has to initiate a request for the transmission of information. Notifications are the cornerstone of *push-based* information delivery via mobile phones as they allow applications to harness the opportunity of steering users' attention toward the delivered information in order to maximize its effectiveness. Indeed, mobile notifications are presented in a unified fashion by almost all mobile operating systems. Usually, in the current implementations, notifications from all applications are listed on the phone's lock screen as well as in a notification bar located at the top of a phone's screen. In order to provide a brief summary of the delivered information to the users, they present a brief summary including the identity of the sender, a short description of the content of the notifications or the event that trigger them, and time of delivery.

Mobile notifications are triggered by humans as well as machines. The former are triggered by recipient's social connections generally through chat and email applications for instantiating communication between two or more persons, whereas, in the latter case, messages are generated in an automatic fashion by the system processes or the native applications, such as system monitoring utilities, scheduled reminders and promotional advertisements. Nowadays, since a variety of sophisticated sensors are embedded in phones, the data generated therefrom is exploited by applications for not just improving usability but also for proactively signaling users (usually through notifications) about the occurrence of events that are associated with their context. Some common example of context-based notifications are collocation-based advertisements [1, 34] and context-based suggestions [57, 102].

Figure 1.1: Information delivery through push notifications on mobile phones.

1.2   ISSUES WITH MOBILE NOTIFICATIONS

Push-based mobile notifications were introduced in mobile phones to keep users free from constantly checking for (i.e., pulling) new information, as they signal users on the availability of any newly arrived information. However, people receive numerous notifications arriving autonomously at anytime during the day through their mobile apps [75, 98]. Such services provide a clear benefit to users as they facilitate task switching and keep users aware of a number of information channels in an effortless manner. However, at the same time, these notifications are often triggered at inappropriate times as they do not have any knowledge about recipients' situation. Psychological studies have found that notifications arriving at unsuitable moments often become a cause of disruption for the on-going task [80, 111]. Notifications delivered at the wrong time can adversely affect the current primary task's execution [8, 21, 22, 81] and users' affective state [2, 7].

A previous study has found that in order to not miss any newly available information that is deemed important, people are willing to tolerate mobile interruptions [52]. However, various mobile applications exploit users' attention by triggering a large number of potentially unwanted notifications [75]. At the same time, some studies have demonstrated that not all notifications are accepted by users as their receptivity is dependent on the type and sender of information being delivered [75, 77]. For this reason, notifications that are uninteresting and irrelevant are mostly dismissed (i.e., swipe away without clicking) by users [30, 98]. Some examples include new game invites, app updates, predictive suggestions by recommendation systems, and marketing messages. Furthermore, continuous trigger of notifications at inappropriate time and context becomes a potential cause of annoyance for users. This might lead them to uninstall the corresponding applications [28, 98]. In order to provide an in-depth overview of these issues, we devote a part of the article to the discussion of issues concerning the cost of interruptions for both users and the service providers.

1.3 SOLUTIONS FOR INTERRUPTIBILITY MANAGEMENT

Interruptibility management has attracted the interest of human-computer interaction researchers well before the advent of mobile devices. However, interruptions received on the desktop have very specific characteristics. In fact, because of their very nature desktops are situated in a constant environment and a user's physical context (such as surrounding people, location and physical activity) does not always change while they are interacting with desktops, whereas mobile devices are carried by users almost everywhere, which makes the physical context of these devices very dynamic. Therefore, interruptibility management for desktop environments is in a sense less complex.

To address the challenge of understanding and modeling users' complex behavior in terms of interaction with mobile notifications, studies have focused on exploiting the contextual information (captured through mobile sensors) [77, 98]. Indeed, the interaction of users with notifications is conditional on various contextual dimensions, which might be partially captured by means of the sensors embedded in the phones. However, by monitoring users' interaction with notifications, researchers have been developing a variety of interruptibility management mechanisms for both desktop and mobile environments. More specifically, these studies have been focused on:

(i) understanding factors associated with users' interruptibility and receptivity to notifications;

(ii) inferring opportune moments and contextual conditions for notification delivery; and

(iii) identifying and filtering notifications that are deemed uninteresting or irrelevant for users.

Consequently, all studies have focused on various challenges concerning the understanding and learning of users' behavioral patterns in terms of interactions with notifications. However, the characterization of users interruptibility for delivering mobile notifications is still an open problem. We believe that there is still a considerable scope for improvement, for example by exploiting other physical, social, and cognitive factors for modeling users' notification interaction behavior. Therefore, in this work we discuss previous efforts and the state of the art in interruptibility management system design for both desktop and mobile environments separately. We then present a critical discussion of these approaches and highlight the current challenges and opportunities in the area that must be investigated to build intelligent mechanisms that could effectively trigger the right information in a given context.

In this book we present an overview of the current state of the art in the area of intelligent notification mechanisms that rely on the awareness of users' context and preferences. The lecture has been designed for a broad readership. It only assumes a very basic knowledge of human-computer interaction and ubiquitous computing concepts. The chapters cover both theoretical and practical aspects related to the implementation of system for intelligent notification management.

CHAPTER 2

Interruptions

In this chapter we discuss different definitions of interruption, provide a possible classification of interruptions, and give an overview of the sources of interruptions.

2.1 DEFINITIONS OF INTERRUPTIONS FROM DIFFERENT RESEARCH FIELDS

The concept of *interruption* has been defined and interpreted in different ways by researchers working in different communities. For example, in linguistics, an interruption has been defined as:

"*A piece of discourse that breaks the flow of the preceding discourse. An interruption is in some way distinct from the rest of the preceding discourse; after the break for the interruption, the discourse returns to the interrupted piece of discourse [38].*"

An interruption has been defined as follows in psychology:

"*An event that breaks the coherence of an ongoing task and blocks its further flow. However, people can resume the primary task that has been interrupted once the interruption is removed [69].*"

On the other hand, in computer science, it has been defined as:

"*An event prompting transition and reallocation of attention focus from a task to the notification [67].*"

In the following sections we will first identify various types of interruptions in order to derive a definition that will be used throughout this lecture.

2.2 TYPES OF INTERRUPTIONS

Interruptions are pervasive in nature. As described by Miyata and Norman [80], in our everyday life we receive numerous interruptions that are both internal and external. An overview of the definitions of these two types of interruptions are as follows.

2.2.1 INTERNAL INTERRUPTIONS

An *internal* interruption occurs due to our own background thought process. More specifically, such interruptions are actions performed by people themselves that lead to break their focus of conscious attention to perform another activity without being prompted by an external event. Such interruptions can also be referred to as self-interruptions as people direct their attention to a different task because of their spontaneous cognitive events.

Let us imagine that a student is working on an assignment and she suddenly remembers that today is her friend's birthday. She takes a break for the assignment and launches the Facebook app on her mobile to wish happy birthday to her friend. Once she posted a birthday wish on her friend's wall, she comes across a post on her timeline that she finds interesting. Eventually, she goes back to work on her assignment, but it took her a few minutes to recall where she was with the assignment.

2.2.2 EXTERNAL INTERRUPTIONS

An *external* interruption is caused by the arrival of an event around a user that is outside of your control. Communication through computing devices or in-person is the fundamental source of external interruptions. As shown by Dabbish et al., external interruptions are likely to increase the tendency of self-interruption for the users [24]. In other words, this demonstrates that a user tends to recall other tasks or events on being interrupted by an external source.

External interruptions can be further divided into two classes depending on the relevance of sources with the primary task.

- **Implicit Interruptions**: These are the interruptions that arrive from some process of a primary task. Such interruptions are mostly relevant to the current task, such as an error message from the application which a user is interacting with.

- **Explicit Interruptions**: These are the interruptions that arrive from a process that does not belong to the ongoing task. Such an interruption might cause an expected task switch from the current activity to a newly introduced activity, such as the arrival of a chat message while a person is interacting with a text editor application.

2.3 DEFINITION OF INTERRUPTIONS IN CONTEXT OF THIS LECTURE

Previous theories about interruptions in different domains enable us to gain a deeper insights about the meaning of interruptions. Starting from this previous work from different communities, we first derive a generic theoretical construct for providing a definition of *interruptions* and framing the problem of *interruptibility*. The objective of this lecture is to focus on external interruptions through mobile devices without considering internal interruptions. Therefore, we define an interruption as *an unanticipated event that comes through a communication medium and has a potential to instigate a task switch and break the flow of the primary activity by capturing users' attention through visual, auditory, or haptic cues.*

Some examples of interruptions that are covered with the above definition are as follows.

- **Scenario 1.** Arrival of a notification on a mobile phone during a meeting. Here, the interruption caused by the notification could adversely affect the ongoing meeting. In such a scenario, if we could defer the notification alert until the end of meeting or a break, it could help not only to avoid that interruption, but also to gain attention of the user.

- **Scenario 2.** Notification alert from an app while interacting with another app on phone. In this case, the affect of interruption depends on the level of engagement with the primary app. For instance, if the primary app was just launched, the interruption cost might not be high. In contrast, if the app is used for a critical purpose, such as navigation during driving, the notification could have an extremely high cost of interruption.

- **Scenario 3.** Delivery of a critical notification without an alert. In cases where people use mute mode to control the plethora of notifications interrupting them, it is possible that sometime people miss some important and time critical piece of information. Even though the user was busy in another task, due to the criticality of information the cost of interruption is low.

2.4 SOURCES OF INTERRUPTIONS

Given the focus on external interruptions, in this section we discuss the sources of such interruptions and the ways in which users handle them.

2.4.1 INTERRUPTIONS IN HUMAN-HUMAN DISCOURSE

In a human-human communication environment, when an interaction is initiated by a person (i.e., the speaker), the listener generally gives feedback about the failure or success of the initiated communication [18]. Such feedback is merely a brief reaction through eye contacts, head nods or voice response. This acknowledgement informs the speaker whether the listener is welcoming and attending the communication or not.

As suggested by Clark and Schaefer [18], "*[f]or people to contribute to discourse, they must utter the right sentence at the right time.*" The physical presence of a person (listener) enables the speaker to determine the right moment to initiate communication. However, the speaker does not usually have a reasonable understanding of the listener's cognitive situation. Consequently, people sometimes initiate communication at wrong moments, which often results in causing interruptions.

In his book [17], Clark has suggested that a listener responds to such interruptions in four possible ways:

(i) responding to interruptions immediately;

(ii) acknowledging and agreeing to handle it later;

(iii) explicitly refusing to handle it by notifying the speaker; and

(iv) implicitly refusing to handle it by not providing an acknowledgement to the interruption.

On the other hand, once the conversation begins, it does not always go error-free. As argued by Sacks et al. [97], the turn-taking during the conversation is itself vulnerable to error. When a person speaks, the other person listens, but sometimes, unintentionally, both start talking simultaneously and people try to coordinate their conversation in an appropriate way. However, this can be categorized as an implicit interruption (discussed earlier in this section), which is not in the scope of this lecture.

2.4.2   INTERRUPTIONS IN HUMAN-COMPUTER INTERACTION

Personal computing devices such as desktops, laptops, and mobile phones, offer a great value to users by facilitating multifarious, informative, and computational functionalities salient to their daily requirements. However, the provision of a platform that runs multiple applications simultaneously, which are delivering various types of information from different channels, often leads to an environment that distracts users from their primary task.[1] This is due to the fact that the information delivered to the users is often not relevant to their primary task, which leads them to switch their attention from the application in focus toward the application being executed in the background that delivered the information. Moreover, applications leverage notifications in order to trigger alerts that try to gain users' attention toward the delivered information. Therefore, even though people try to ignore all interruptions and continue focusing on their primary task, they still receive cues about the newly delivered information, which might cause information overload [100].

In 1997, by looking at the trend toward the development of "intelligent" computer systems and technologies, McFarlane envisioned the hazards from such intelligent systems competing for users' attention [68]. McFarlane, for the first time, investigated the strategies for counteracting interruptions caused by intelligent computer systems. He built a taxonomy based on theoretical constructs that are relevant to interruptions. This taxonomy identifies the following eight descriptive aspects of human interruption:

(i) *Source of interruption*: who triggered the interruption;

(ii) *Characteristics of the user being interrupted*: receiver's perspective for getting interrupted;

(iii) *Coordination method*: approach used for determining the moment to trigger interruption based on users' response;

(iv) *Meaning of interruption*: what the interruption is about;

---

[1]Here a primary task can be any operation that the user is currently performing, which might or might not include interaction with a computing device.

(v) *Method of expression*: design aspect of the interruption;

(vi) *Channel of conveyance*: medium of receiving the interruption;

(vii) *Human activity changed by interruptions*: internal or external change in the recipient's conscious and physical activity; and

(viii) *Effect of interruption*: impact of interruption on an ongoing task and the user.

Instead of limiting the scope of his work to HCI, McFarlane built this taxonomy from an interdisciplinary perspective drawing upon the theories for human interruption discussed in the literature from many different domains. Each dimension of the taxonomy describes a unique aspect of human interruptibility.

In 1999, Latorella proposed an Interruption Management Stage Model (IMSM) that describes information processing stages on receiving interruptions [61]. This model can be used to study information processing by humans and to identify the effects of interruptions in different stages of information processing. The model was designed with an assumption that recipients are engaged with an ongoing task with which they are familiar and that it can be resumed at any point. The model comprises of three stages.

(i) *Interruption detection*: when a user is engaged with the primary task, a salient alert is required in order to initiate an interruption.

(ii) *Interruption interpretation*: on detection of an interruption, the user's attention is directed toward the interruption for further processing in order to interpret the requirements of the interrupting task.

(iii) *Interruption integration*: in this final stage, the user integrates the interruption with the primary task by immediate or scheduled tasks switching.

In 2002, McFarlane and Latorella investigated users' behavior on receiving interruption alerts (i.e., the first stage of IMSM model) [69]. They argued that user's response to computer generated interruptions is similar to the response to interruptions during human dialogue as proposed by Clark [17]. However, they suggested that Clark only considered the user's response for detected interruptions; indeed, in the human-computer interaction setting these can also go undetected. Therefore, undetected interruptions might represent additional aspects of user response, which should be considered when building interruption management system for computing environments.

McFarlane and Latorella proposed the following five key responses of users to an interruption arriving during the process of human-computer interaction.

(i) *Intentional integration*: the interruption is relevant to the ongoing task and the user integrates it with the ongoing task.

(ii) *Preemptive integration*: the interruption is handled immediately or scheduled for later.

(iii) *Oblivious dismissal*: the interruption goes unnoticed by the user and, thus, it is not performed.

(iv) *Unintentional dismissal*: the interruption is not performed as it is not interpreted to the user.

(v) *Intentional dismissal*: the user explicitly decides not to handle the interruption.

The first two categories of responses suggest that the interruption content is acceptable to the user. However, preemptive integration indicates that the delivered interruption is not relevant to the ongoing task at the moment of its delivery and, thus, it should have arrived after a certain time delay in order to prevent immediate task switching. On the other hand, the responses with a dismissal reflect that either no interruptions or a particular interruption is not at all acceptable to the users in their current situation. Here, oblivious and unintentional dismissals indicate that there is a need to highlight the importance of redesigning the attention cueing (to avoid a notification getting unnoticed) and highlight the summary of the delivered content (to make the interruption salient). Similar to preemptive integration, intentional dismissal indicates the need of scheduling interruption at opportune moments to improve the probability of its acceptance.

CHAPTER 3

Cost of Interruption

Interruptions are an inevitable part of our everyday life as it is hard to get through the entire day without being interrupted. As suggested by Zabelina et al. in [109], people are sensitive to their surroundings and they receive more information through interruptions, which might help them in their everyday tasks and even boost their creativity. This represents a positive aspect of interruptions helping users to effortlessly receive information from different sources. Numerous studies [5, 8, 21, 23] have also demonstrated that interruptions have a detrimental effect on users' memory, emotional and affective states, and ongoing task execution. These findings indicate a negative facet of interruptions when they arrive at inappropriate situations.

In this chapter we discuss the cost associated with the arrival of interruptions at inopportune moments. The summary of the findings from studies investigating the impact of interruptions is summarized in Table 3.1.

3.1 IMPACT ON MEMORY

In 1927, Zeigarnik performed a classic psychological study [110] (as cited in [6]) with the goal of examining the mechanisms of retrospective remembering with and without interruptions. In this study the participants were given a series of practical tasks, for instance sketching a vase and putting beads on a string. Some tasks were interrupted and others were carried out without any interruption. Tasks could be performed in any order by participants. It was possible to switch to another task without completing the ongoing task, which could be taken up later. On the completion of all tasks, they were asked to do a recall test. The results of this study demonstrate that people can recall the content of interrupted tasks with greater accuracy compared to the case of uninterrupted tasks. This indicates that people have selective memory associated with the interruptions they receive. Such observed behavior is referred to as the Zeigarnik Effect.

Although the Zeigarnik Effect suggests that interruptions are useful for retrospective memory, many other applied studies have argued that interruptions have an adverse impact on memory [5, 21, 27, 36]. In particular, Dix et al. argued that humans can memorize only a limited list of tasks they have been engaged in due to the nature of their cognitive abilities [27]. Moreover, they suggested that if interrupted during a task, humans are likely to lose track of what they were doing. Following these suggestions, Edwards and Gronlund [5] conducted an experiment to investigate the memory representation for the primary task after handling an interruption. Their study was orthogonal to the Zeigarnik Effect experiment as in the latter participants were not asked to resume or recall where they left the primary task on arrival of the interruption.

Table 3.1: Studies in the area of understanding the cost of interruptions

Study Type	Key Findings
Impact on Memory	1. People have selective memory associated to the interruptions they receive [110]
	2. People possess a stronger memory representation of an uninterrupted task as compared to an interrupted task [5, 36]
Relationship with Ongoing Task	1. Interruptions could have an adverse effect on the completion time and errors made while performing a complex computing task compared to a simple task [36, 55]
	2. People perceive less disruption if the interruption is highly relevant to the current task [23]
	3. Amount of disruption perceived also linked to the mental load of a user on the arrival of an interruption [8, 9]
	4. People perceive varying level of disruption while performing different sub-tasks [21, 22]
Relationship with Users' Emotional State	1. People's emotion and well-being are negatively impacted by interruptions [111]
	2. People experience annoyance and anxiousness on arrival of an interruption [2, 9]
	3. Interruptions coming from mobile phones cause lack of attention and hyperactivity symptoms in users [56]
Impact on User Experience	1. Complex interfaces make it difficult for users to handle interruptions [55]

In particular, participants were given a task comprising of ten items and an interruption was delivered after they completed the fifth item. The interruption was in the form of a message indicating that the first phase of the experiment was over. This message was displayed on the screen of participants' primary task and the same message was orally reiterated by the experimenter to ensure the interruption was received by them. Through their experiment, Edwards and Gronlund showed that people tend to need a certain amount of time before resuming back to the primary task after an interruption. Moreover, they demonstrated that people possess a stronger memory representation of an uninterrupted task as compared to an interrupted task.

3.2   IMPACT ON ONGOING TASK PERFORMANCE

In 1981, Kreifeldt and McCarthy [55] argued that interruptions could have an adverse effect on the completion time and errors made while performing a computing task. They demonstrated that people perceive more disruption and become more prone to make errors on getting interrupted while performing a complex task compared to a simple task. Later in 1989, through a series of experiments, Gillie and Broadbent [36] investigated the impact on the primary task by three aspects of an interruption: (i) length; (ii) similarity with the primary task; and (iii) the action required to handle it. Their results show that people feel distracted when interruptions share characteristics with the ongoing task or if they are limited to complex tasks but the length of an interruption does not make it disruptive. However, these findings are not aligned with those of Czerwinski et al. [23] who demonstrated that people perceive less disruption if the interruption is highly relevant to the current task.

In [8], Bailey at al. studied whether the performance of an ongoing task is influenced by interruptions. Their experiment utilized six types of web-based tasks: addition of numbers; counting of items; comprehension of images; comprehension of written text; registration; and selection. Participants were interrupted when they were approximately in the midway to completion of each task. They were presented with a news report or an investment decision as an interruption. Their findings show that: (i) people perform interrupted tasks slower compared to non-interrupted tasks and (ii) the amount of disruption perceived depends on the type of ongoing task. Later, in another study [9], Bailey's et al. demonstrated that the amount of disruption perceived also depends on the mental load of a user on the arrival of an interruption.

Czerwinski et al. [22] studied the impact of interruptions while performing different types of sub-tasks. Their results show that people perceive varying levels of disruption while performing different sub-tasks. They proposed that deferring interruptions until a new subtask is detected could also reduce the perceived disruption. These findings were extended by Cutrell et al. [21] to investigate the effects of instant messaging on different types of computing tasks. They found that the perceived disruptiveness is higher when users are engaged with tasks that require their attention.

3.3   IMPACT ON USERS' EMOTIONAL STATE

In [111], Zijlstra et al. for the first time studied the effect of interruptions on users' psychological state. They investigated whether interruptions produce an adverse effect on users' emotions and well-being, and raise their activeness level. They conducted a series of experiments by creating a simulated office environment for performing realistic text editing tasks. Their findings suggest that users' emotion and well-being are negatively impacted by interruptions, but they do not affect the activeness level.

In 2001, Bailey at al. investigated the effects of interruption on users' annoyance and anxiety levels [9]. Through a series of experiments, they demonstrated that people experience annoyance on arrival of an interruption. The annoyance level experienced by users depends on the type of ongoing task, but not on the type of the interruption task. They also show that the increase in users' anxiety level is higher when they receive interruptions during a primary task as compared to the arrival of interruption on completion of the primary task.

In another study [2], Adamczyk and Bailey investigated the impact on users' emotional state by interruptions arriving at particular times during task execution. In their experiment, participants were asked to perform tasks (such as text editing, searching, and watching video) and a periodic news alert was triggered as an interruption. Their findings show that users experience annoyance and frustration on receiving interruptions. Moreover, interruptions arriving at different moments have a varying impact on users' emotional state.

In a study concerning mobile notifications [56], Kushlev et al. investigated whether interruptions coming from mobile phones cause lack of attention and symptoms linked to Attention Deficit Hyperactivity Disorder (ADHD). They asked participants to maximize interruptions by turning on their phones' notification alerts and trying to mostly be within the reach of their phone. Later, participants were asked to minimize interruptions by turning off their phones' notification alerts and trying to stay away from their phones. Their results show that people reported higher levels of hyperactivity and distraction during the first phase of the experiments. This suggests that by simply adjusting existing phone settings people can reduce inattention and hyperactivity levels.

3.4   IMPACT ON USER EXPERIENCE

"The most profound technologies are those that disappear. They weave themselves into the fabric of everyday life until they are indistinguishable from it."—Mark Weiser [106].

The above quote from Mark Weiser's seminal paper [106] captures and summarizes his vision for ubiquitous computing. His goal was to design an environment with embedded unobtrusive computing and communication capabilities that can blend with users' day-to-day life. The two key aspects of his vision were: (i) effective use of the environment in order to fuse technology with it and (ii) making the technology disappear in the environment.

The second aspect of his vision focuses on the user experience and suggests making technology disappear from the user's consciousness. Another classic paper of Weiser [107] describes the disappearing technologies as *calm computing*. In this paper, the authors suggested that the technology should enable the seamless provision of information to users without demanding their focus and attention. However, interruptions have the potential to take mobile technology and Mark Weiser's vision far apart because they not only create potential information overload but also demand user attention.

As suggested by Mark Weiser [106], technology should be "transparent" to users so that they should not notice that they are interacting with computing devices. In one of the first works in this area, Kreifeldt and McCarthy investigated the design of different user interfaces in order to reduce the effects of interruptions [55]. Their results suggest that interaction design plays a key role in affecting users' ability to successfully resume the interrupted task. More specifically, their findings suggest that complex interfaces make it difficult to handle interruptions.

3.5   INDIVIDUAL DIFFERENCES IN PERCEIVING DISRUPTION FROM INTERRUPTIONS

As defined earlier, interruptions are unanticipated events that fragment the flow of execution of an ongoing task by demanding users to switch their attention to interrupting tasks. This leads to a multitasking environment for users. Humans naturally have skills to handle interruptions and adapt to a multitasking environment; however, they do show individual differences in their ability to perform tasks in such setting [4, 12, 54, 105].

In 1953, Atkinson investigated the role of people's motivation to recall the completed and interrupted tasks [4]. The study demonstrates that highly motivated users are likely to recall interrupted tasks better as compared to recalling uninterrupted tasks. On the other hand, less motivated people show a tendency to recall completed tasks better than recalling the interrupted task. Overall, his findings suggest that the ability to recall the primary task after handling interruptions varies across people. A few years later, similar findings were reported by Bernard Weiner [105].

Joslyn and Hunt proposed "The Puzzle Game"—an empirically validated test that can quantify the performance of users for multitasking [54]. Through a series of experiments, the authors showed that the ability to make rapid decisions for task switching is not the same for everyone. They suggested that people who are good in making quick decisions can be identified through testing their psychological characteristics, which can be captured by their puzzle game.

Brause and Wickens in [12] investigated the individual differences in sharing time between tasks in a multitasking environment. Their analysis show that there are differences in time-sharing ability of individuals; these are linked to their potential to process information. Moreover, their findings suggest that people have different strategies for time-sharing, which introduces differences in individuals' multitasking ability.

Moreover, studies also have demonstrated that people show significant differences in their cognitive style with respect to multitasking [13, 47]. In particular, these tudies have demonstrated that the performance in carrying out an interrupted task is affected by users' anxiety [47] and arousal [13] levels. However, since the levels of anxiety and arousal vary across people, different levels of disruption are perceived by them through the arrival of interruptions at inopportune moments [13, 47].

3.6   SUMMARY

In this chapter we have given an overview of the a large body of work that shows that, even though notifications are extremely beneficial to the users, they still are a cause of potential disruption as they often require users' attention at inopportune moments. In fact, previous studies have found that interruptions at inopportune moments can adversely affect task-related memory, lead to high task error rate, impact the emotional and affective state of the user, and hinder their experience of interacting with computing devices. In the next chapters, we will present the solutions proposed by numerous studies to address the issue of delivering the right information at the right time.

CHAPTER 4

An Overview of Interruptibility Management

4.1 DEFINITION

Information delivered through computing devices often arrives at inopportune moments. This might adversely affect our ongoing tasks and psychological states. In order to address the issue of delivering the right information at the right time, researchers and practitioners have proposed various designs for building effective *interruption management* systems. Interruption management is a process that combines technology, practices and policies to build solutions for controlling interruptions from seeking users' attention at inopportune moments. The key objective of an interruptibility management system is to help users to effectively perform their primary task and make computing devices *calm* by unobtrusively mediating interruptions [48].

Figure 4.1 presents the architecture of an interruptibility management mechanism entailed in an app. The interruptibility management mechanism handles interruptions from both local and remote notifications. Local notifications are generated through system process and other native apps, and the remote notifications are triggered via the back-end server supporting the app. The interruptibility management system entails an *interruptibility model*, which is used to drive the delivery of notifications at times that are considered opportune given, for example, the current context of the user and certain characteristics associated with the information to be dispatched.

As shown in Figure 4.2, a high level overview of the process to build interruptibility models consists of three steps: data collection, model construction, and interruptibility prediction. These steps are defined as follows.

- **Data collection.** The interruptibility management system monitors the interaction of users with notifications they receive on their device (e.e., a mobile phone). Along with the notification interaction data, it also collects users' contextual information that is associated with them. The type of information that is collected depends on the type of interruptibility model itself. There are different types of datasources that can be used to build these models as discussed later in this section.

- **Model construction.** The collected data is analyzed to derive some features that are considered to be informative for the learning process of the predictive model. This process is called *feature extraction*. The extracted features are then used to build the predictive model,

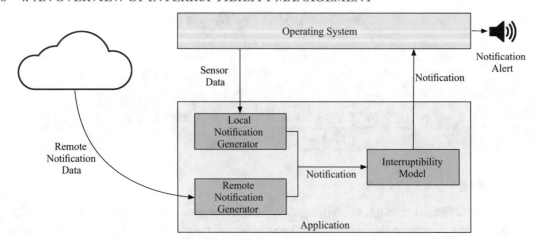

Figure 4.1: Architecture of an interruptibility management mechanism.

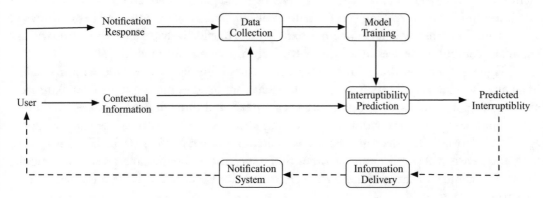

Figure 4.2: Process of building an interruptibility model. The process can happen *offline* (i.e., the responses to notifications are collected using lab or in-the-wild studies for data collection without the presence of an interruptibility management system) or *online* (i.e., the model training is refined by considering the responses to the notifications sent by the means of an interruptibility management system itself).

which is generally based on a machine learning algorithm. The features are given as input to the machine learning algorithm along with the corresponding users' interaction behavior. The model is trained to learn patterns in the features that can be used for predicting the corresponding interaction. Also, depending on the interruptibility management system's design, the model is sometimes retrained after a certain period of time to enable the system to adapt to any changes in the user's behavior over time.

- **Interruptibility prediction.** Once the model is trained, it can be used to make predictions by supplying the contextual data as input. The contextual data is used to extract features (as discussed in the previous step), which are then exploited by the model to predict interruptibility of a user in a given situation.

4.2   KEY DIMENSIONS FOR THE DESIGN OF INTERRUPTIBILITY MANAGEMENT SYSTEMS

There are two key dimensions for the design of interruptibility management systems: (i) delivering right information at the right time and (ii) delivering information through the right medium. Most of the existing work in this area (discussed later in this book) has focused on the former dimension. The latter dimension has instead become the focus of the research community in the recent years.

4.2.1   DELIVERING RIGHT INFORMATION AT THE RIGHT TIME

An interruptibility management system should be designed considering the following two aspects: the right *type* of information and the right *delivery time*. More specifically, *right time* indicates that the information should be delivered when the user is likely to quickly attend it, whereas, *right information* refers to delivering the information that is considered relevant to the user in the current context. In the area of interruptibility management, right time to deliver information is referred as *attentiveness*, and *receptivity* is the term used to indicate the delivery of right information. We will now focus on the two key aspects of interruptibility: attentiveness and receptivity.

In [91], *attentiveness* is defined as the amount of attention paid by users toward their computing device for a newly available interruption task. However, attentiveness does not consider the response of the user to the interruption, which can be either positive (i.e., the interruption is accepted) or negative (i.e., the interruption is dismissed). On attending an interruption users get subtle clues about different features of interruptions, such as a brief description of the content, sender, and urgency of the interrupting task, which helps them to decide whether to click or decline those interruptions.

On the other hand, *receptivity* is defined as the process of making a decision about the way in which the user is willing to respond to an interruption by analyzing its clues. In [30], Fischer et al. argue that users' receptivity to an interruption not only encompasses their reaction to that specific interruption but also their subjective experience of it. However, users' receptivity varies with the context as it accounts for their negotiation in handling interruptions in different contexts.

For a practical point of view, in Figure 4.3, we present a scenario in order to understand attentiveness and receptivity from a system point of view. Attentiveness can be quantified as the time taken by the user to see a notification after its arrival, whereas, receptivity comes into act

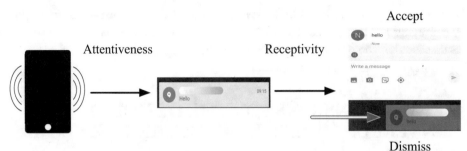

Figure 4.3: The process of users' interaction with a notification. Attentiveness is the time taken by the user to see a notification after its arrival. Receptivity is the user's decision to accept or dismiss a notification.

only when the user is already attended a notification, and it is defined as the user's decision to accept or dismiss a notification.

4.2.2   DELIVERING INFORMATION THROUGH THE RIGHT MEDIUM

Another dimension of an interruptibility management system is the choice of the *right medium* for delivering information. This dimension is applicable only to information delivered by *cross-platform applications* in a *multi-device environment*. Here, a multi-device environment refers to the situation when users are surrounded by a variety of computing devices. This is very common these days as people usually have access to two or more computing devices such as laptops, computers, mobile phones, tablets, and smartwatches, etc. This creates an environment where the user is connected to numerous information channels through different devices. Examples include accessing an email app from laptops and playing games on mobile phones. However, sometimes people have the same source of information linked on multiple devices they own. An example is a chat app client installed in different devices owned by users, such as their mobile phones, smartwatches, laptop and so on. The applications that can be installed on multiple devices are usually referred to as *cross-platform* apps.

In general, cross-platform apps facilitate users with an ease of accessing information from any device they are connected on. Although this sounds very useful, it also makes the interruptions inevitable and, thus, increases the amount of perceived disruption. The issue is exaggerated when the information is handled by the user on a certain device but it is still delivered on other devices, which is due to the communication delay of a cross-platform app (e.g., receiving a Skype call on mobile phone and tablet, even though the user has accepted the call on their laptop). Therefore, in today's world it has become fundamental for an interruptibility management system to address the issue of delivering information through the right medium.

CHAPTER 5

Interruptibility Management in Desktop Environments

Interruptibility management has attracted the interest of HCI researchers well before the advent of mobile devices. However, interruptions received on the desktop have very specific characteristics. In fact, because of their very nature desktops are situated in a constant environment and a user's physical context (such as surrounding people, location, and physical activity) does not always change while they are interacting with desktops. This indicates that the potential factors associated with the interruptibility of users are limited and can mostly be detected by monitoring their interaction with desktops. Indeed, there could be other factors in the surrounding of users that change, but not frequently, and have a significant impact on their interruptibility.

In contrast, mobile devices are carried by users almost everywhere, which makes the physical context, in which these devices are used, very dynamic. At the same time, today's mobile devices carry a plethora of apps that overload the users with a flood of notifications arriving at anytime regardless of the situation, which causes much severe disruption compared to desktop notifications. This dynamic usage of mobile devices demonstrates that the interruptibility of users in such an environment is not just associated with the primary task they are performing but also numerous other aspects of their surroundings. In general, interruptibility management for desktop environments is in a sense less complex compared to mobile environments.

Indeed, there have been numerous approaches proposed for designing an interruptibility management system for desktops [11, 43, 44, 46, 51, 53]. These approaches can be categorized based on the obtained inputs for building their models to predict interruptibility. As a result, there are two types of approaches taken for designing interruptibility management systems for desktops: (i) by exploiting task phases and (ii) by using sensor data.

5.1 INTERRUPTIBILITY MANAGEMENT BY EXPLOITING TASK-RELATED INFORMATION

In 1986, Miyata and Norman argued that people are less prone to perceive interruptions as disruptive in some phases of a task compared to other phases [80]. This suggestion was later investigated by Czerwinski et al. [23] in 2000 and Cutrell et al. [21] in 2001. Both studies confirm the insights of Miyata and Norman that perceived disruption varies when an interruption arrives at different phases of the task. Moreover, Czerwinski et al. [23] found that interruptions are perceived as less disruptive when they arrive after the primary task is completed.

On the other hand, in [21] Cutrell et al. demonstrated that interruptions arriving at the beginning of a primary task are perceived as less disruptive than interruptions occurring at other phases of a task execution. They suggested that interruptions should be deferred until a user is switching tasks, rather than delivering them immediately. They designed an interface for instant messaging that constantly monitors user actions and infers the different phases of the task completion level as well as the moment when the users switch from one task to another. They also suggested that this information can be used to deliver instant messages at opportune moments.

Horvitz et al. devised the term "bounded deferral" [42], which utilizes the concept of deferring the interruption if the user is busy and determines the time until which the interruption should be held from being delivered in order to minimize the disruption cost without losing the value of information due to the delay. The interruption is deferred until a maximum amount of time that is pre-specified by users (known as maximal deferral time) and after this maximal deferral time has elapsed, the alert is triggered immediately even if they still remain busy. They examined the *busy* vs. *free* states for 113 users over a period of two consecutive days (i.e., a day for each situation). Participants were provided with a "Busy Context" tool that allows them to define when they were busy or free. The analysis of the data showed that users switch from busy to free situations in approximately 2 min. Moreover, they demonstrated that medium and low urgency emails can be deferred for 3 and 4 min, respectively. They suggested that bounded-deferral policies can reduce the level of interruption while allowing users to be aware of important information.

As discussed earlier in Chapter 3, Adamczyk et al. demonstrated through a controlled experiment that the breakpoints within a task are the opportune moments to deliver interruptions [2]. Later, in [3] they proposed a system that can automatically infer the breakpoints in tasks and exploit this information to deliver interruptions. In order to build this system, they leveraged the findings of [10, 41, 82], which suggested that users' mental workload has a statistical correlation with the size of the pupil. They validated this in a human-computer interaction environment by showing pupillary response aligns with the changes in mental workload [48, 53]. They built a system that uses a head-mounted eye-tracker for measuring users' pupil size. Finally, they showed that their system was able to infer mental workload for route planning and document editing tasks with an average error of 3% and 2%, respectively.

In 2007, Iqbal and Bailey investigated the feasibility of inferring different classes of breakpoints (coarse, medium, and fine) while a task is performed without using any supplementary hardware resources [50]. Data was collected from several participants in the form of event logs and screen interaction videos while they were performing the task. They recruited observers to watch the participants performing the task in order to identify and label breakpoints, their type and explanations for their choice. Using the suggestions reported by Fogarty et al. [32, 33], they identified a series of features (such as "switched to another document," "closed an application," "completed scroll," and so on) by analyzing observers' explanations for breakpoints and event logs. Based on these features they constructed statistical models, which were able to predict each type of breakpoints with an average accuracy of 69–87% (i.e., the percentage of correctly identified breakpoints over the total predicted breakpoints). For each type of breakpoint, different features were computed and also different models were constructed.

Previously, Iqbal and Bailey [49] investigated whether the cost of interruption can be predicted by exploiting the characteristics of task structure. They posited that the cost of interruption is measured only by the resumption lag (i.e., the time taken by users to resume back to their primary tasks after handling an interruption). More specifically, they tried to test whether interruptions arriving at boundaries of subtasks have low cost of disruption, by using the task structure rather than using head-mounted cameras to measure pupil size as in [50]. Here, *task structure* indicates the decomposition of a task into a sequence of subtasks. The characteristics of the task structure refers to the level at which the task is broken down into subtasks, the type and the mental load associated with these subtasks [15]. They evaluated their approach by conducting experiments on a set of primary tasks. In order to collect the data for estimating the resumption lag on receiving interruptions, participants were interrupted at various boundaries of the task execution. Finally, they constructed a statistical model that could predict the cost of interruption with a 56–77% accuracy for all tasks.

5.2   INTERRUPTIBILITY MANAGEMENT BY USING CONTEXTUAL DATA

In this section we present the studies that exploited contextual data for designing interruptibility management systems for desktop systems. We divide the studies into two groups based on the type of analysis performed.

(i) **Offline Analysis**—this group consists of studies that present an *offline* construction and training of models to identify opportune moments for triggering interruptions.

(ii) **On-the-fly Inference**—this group comprise of studies demonstrating the design of systems based on on-the-fly inference of interruptibility.

5.2.1   OFFLINE ANALYSIS

In 1999, Horvitz et al. argued that the cost of an interruption depends on the nature of the interruption and on the user's current task and focus of attention [43]. To investigate this issue they started the Lumiere project [43] with the goal of building a system for attention-sensitive notification delivery based on a probabilistic model. Their system aims at making predictions about the user's interruptibility by considering two separate models for prediction: (i) users' attention and (ii) cost of deferring an interruption. In order to construct a model for users' attention they employed the Wizard of Oz technique to conduct a series of experiments.[1] The data obtained from these experiments was later fed to a Bayesian network [35] to obtain a probabilistic distribution for the degree of the user's attention. The constructed Bayesian models were used to predict the probability distribution over a user's focus of attention.

For the second part of their system, they used a support vector machine [20] for building a model to predict the cost of deferring an interruption by exploiting factors related to the criticality of messages. These features include: sender, recipient, time criticality, use of past or future tense, coordination, personal requests, embedded keywords for importance, length of message, presence of attachment, and time of the day. In order to obtain these features, they relied on natural language processing for extracting information from the content of notifications. To evaluate this, they trained the model with 1500 emails comprising of approximately equal sets of low and high priority email messages. The results, obtained from the dataset of 250 high- and 250 low-priority messages, show that the model could predict the priority (in terms of high and low) with a very high accuracy.

In 2003, Hudson et al. argued that opportune moments to interrupt people can be better predicted by considering a wide range of context information. For the first time, in [46] Hudson et al. explored the possibility of detecting estimators of human interruptibility by using sensors in order to enhance computer mediated communications in work settings. They conducted a study using a Wizard of Oz approach that enabled them to simulate sensing functionalities without implementing actual sensors. The study was based on the audio and video recordings of the working environment of participants. The data was collected during their working hours for the duration of 14–22 working days for each participant. These recordings were then manually analyzed by coders to point out the moments for 23 events, which can be practically sensed. These events included information about the users, people around them, and the surrounding environment.

During the period of data collection, participants were also prompted to report their interruptibility on a 5-point Likert scale. These prompts were triggered randomly with an average of two prompts per hour. Overall, they collected data for a total of 672 prompts, which were responded by the participants. Using the collected data they constructed a series of predictive models that achieved a level of accuracy in the 75–80% range. Their study provided evidence that sensor information can effectively be used to estimate human interruptibility.

Later, in [31], the authors improved these models by exploiting additional sensor readings. Through a series of experiments, they demonstrated that their models attained a significant improvement in terms of robustness. For the evaluation of these models, the authors conducted experiments in real-world scenarios [33]. In particular, the authors used real sensors correspond-

---

[1]The so-called Wizard of Oz experimental technique involves a user being observed while operating a system whose functionalities are simulated by a wizard (i.e., a hidden observer) [25]. Here, the system is made to appear as a real one as the wizard, who is located behind the system or remotely connected to the system, fakes the effect of all functions that are not implemented in the system. This makes people believe that they are interacting with a fully functional system and, thus, enables the investigators to monitor the responses of users to their system without actually implementing it.

ing to the simulated ones from their earlier work and deployed them in an office environment with a much more diverse group of workers [33]. The results from the analysis of the gathered data show that the model has an accuracy of 87%, which is significantly better than that achieved in their prior work.

They also examined the amount of data required to create these models in order to achieve a reliable estimate of interruptibility. Their findings suggest that information collected from other people can be used for an initial training of the models and still be able to achieve a reasonable accuracy. The proposed solution relies on a continuous adaption of the model over time according to the specific behavior of the user. They demonstrate that their system is able to adapt well to different types of office workers.

5.2.2   ON-THE-FLY INFERENCE

All of the studies discussed above focus on an *offline* construction and training of models to identify opportune moments for triggering interruptions. Lilsys [11] and BusyBody [44] were the first attempts at designing solutions for on-the-fly inference of interruptibility. Both systems were constructed with purpose-built custom hardware designed for these research projects and focused on interruptions in an office environment.

Lilsys [11] uses ambient sensors to detect user's unavailability for telecommunication. The system consisted of sensors including sound and motion sensors, phone and door usage inference through the attached wires and keyboard/mouse activity inference. Moreover, it allowed users to manually register their unavailability (if they wanted to) and turn on/off the sensing whenever they wanted. Data is collected passively and an inference about the user's presence and availability is made on detecting a change in any sensor event. Lilsys uses the data from phone, keyboard, mouse, motion and sound detectors for predicting the user's presence. At the same time, the unavailability is predicted by exploiting the data from sound, phone, and door sensors. Both predictors are based on a simple Decision Tree model. The system was installed in an office for around seven months. Participants reported the qualitative improvement in the interruptions but not much reduction in the quantity of interruptions.

BusyBody [44] is based on an initial training phase during which the system asks users about their interruptibility at random times and it also continuously logs the stream of desktop events, meeting status and conversations. Then, on completion of the training phase it uses the collected information to build predictive models based on Bayesian networks for inferring the cost of interrupting users (i.e., the level of perceived disruption) in real time. Through a small-scale study, the authors demonstrated that BusyBody is able to make predictions about the cost of interrupting users with an accuracy of 70–87%.

Both Lilsys and BusyBody represent valuable applications of machine learning algorithms for exploiting the contextual information to predict interruptibility. Similarly, Iqbal and Bailey proposed OASIS [51]—a system that detects the breakpoints in users' activity independent of any task by exploiting the streams of application events and user interaction with a computer.

This information is used to determine the notification scheduling policies on-the-fly and to deliver notifications accordingly.

5.3   SUMMARY

There have been numerous approaches proposed for designing an interruptibility management system for desktops. In Table 5.1, we present the summary of the literature in this area. Previous approaches toward designing interruptibility management system for desktops are categorized into two groups based on the data used for building interruptibility models: (i) by exploiting task related information and (ii) by using sensor data.

Table 5.1: Studies in the area of interruptibility management in desktop environment

Types of Information Used for Interuptibility Prediction	Types of Approach
Task-Related Information	Inferring breakpoints within tasks and exploit this information to deliver interruptions [10, 21, 23, 41, 82]
	Inferring cognitive load through pupillary response and determining the interruptibility accordingly [48, 53]
	Determining the time until which the interruption should be held from being delivered in order to minimize the disruption [42]
	Using system usage features to determine interruptibility [32, 33]
	Predicting the cost of interruption by exploiting the characteristics of task structure [49]
Contextual Data	Predicting the level of users' attention by using features of their interaction with the system [43]
	Predicting interruptibility by using contextual information obtained through simulated sensors [31, 32, 33, 46]
	Using ambient sensors to detect user's unavailability for telecommunication [11]
	Predicting interruptibility by using contextual information including stream of desktop events, meeting status, and conversations [44]

The first group of studies have shown the potential of extracting information about the user's interaction with the primary task and using it for the inference of their interruptibility. The features, about users' interaction with their primary task, exploited in these studies includes: breakpoints within tasks, stage of the task completion, task structure, and cognitive load detected through pupillary response.

The second group of studies have shown the potential of designing sensors to capture contextual information about users that can be exploited for predicting their interruptibility. These studies were mostly conducted by employing a Wizard of Oz technique and recoding users' interaction with notifications. The recordings were analyzed by the coders for simulating the sensors. These studies exploited features related to users' interaction with the system, their surrounding environment, and their routine extracted through calendars.

Even if we spend less and less time interacting with desktop computers, we believe that the studies covered in this chapter provide very important design principles for designing interruptibility management systems independently from the devices themselves. In the next chapters, we will present solutions to address the issue of delivering the right information at the right time in mobile and multi-device environments.

CHAPTER 6

Interruptibility Management in Mobile Environments

When mobile devices first appeared, they were used merely for calling and messaging purposes. Later, with the advent of sensing capabilities, these devices have graduated from calling instruments to intelligent and highly personal devices performing numerous functions salient to users' daily requirements. This has provided opportunities to mobile applications to connect users to different information channels and deliver them updates in real-time about a variety of events, ranging from personal messages to traffic alerts and advertisements. Notifications are at the core of this information awareness, as they use audio, visual, and haptic signals to steer the user's attention toward the newly arrived information.

Studies have found that everyday a much higher number of notifications are triggered on people's mobile phone compared to desktop notifications [75, 90]. As people always carry their mobile phones with them, mobile notifications could arrive at anytime and anywhere, which makes mobile interruptions inevitable and more obnoxious compared to desktop interruptions. In other words, even though mobile notifications are extremely beneficial to the users: however, at the same time, they are a cause of potential disruption, since they often require users' attention at inopportune moments. Thus, managing interruptions in mobile settings has also become a more complex and important task.

In this chapter we focus on the literature of the area of interruptibility management for mobile phones by analyzing the approaches taken in previous studies for designing a solution for managing interruptibility. There are two key dimensions that researchers have focused on while designing interruptibility management systems: (i) delivering right information at the right time and (ii) delivering information through the right medium. Before moving to discuss a variety of solutions proposed to design interruptibility management systems for mobile phones, we will present the findings of studies focused on understanding the experience of users while they receive and interact with mobile notifications. Later in this chapter we will discuss how these findings are incorporated by studies proposing different approaches for designing interruptibility management systems.

6.1   UNDERSTANDING USERS' PERCEPTION TOWARD INTERRUPTIONS CAUSED BY MOBILE NOTIFICATIONS

Even though people report that notifications are disruptive, they still like to continue receiving them in order to keep themselves aware of newly available information automatically instead of manually checking [52]. People tend to use some simple strategies of their own in order to manage interruptions. In a study with several participants [16], Chang and Tang found that people mostly manage interruptibility through the ringer mode of mobile phones. Findings of another study [108] suggest that only a few people tend to change notification settings for individual applications, such as stopping notifications from a specific application. Similar findings were reported by Lopez et al. [65], demonstrating that people tend to stop all notifications from applications that are causing most interruptions by delivering numerous notifications, and, thus, they have to manually check these applications from time to time so that they do not miss any important notifications. The authors of the study also suggested that people want a fine-grained control for how, when and which notifications are delivered to them, which is not present in any mobile platform. However, due to the high number of apps installed on people's phones these days [86], it gets difficult to manually control the delivery of certain types of notifications from each app. At the same time, people's preferences for managing interruptions from mobile notifications change over time as reported by multiple studies [70, 75].

In [98], the authors conducted a large-scale data collection involving around 200 million notifications from 40,000 mobile phone subscribers to investigate the user behavior on receiving mobile notifications. They found that even though a plethora of notifications are triggered on mobile phones (around 60 notifications per day), people handle most notifications within a few minutes from their arrival. Moreover, by analyzing subjective responses of participants (collected through the ESM-based questionnaires), they found that notification triggered by different types (i.e., categories) of apps are assigned different importance by the users. Their findings suggest that notifications comprising information about individuals and events (such as notifications triggered by messaging applications) are given the most importance.

In another study [90], Pielot et al. found that most mobile notifications received by people are about personal communication, i.e., they are triggered by messenger and email applications, which was later confirmed by Mehrotra et al. in their study [75]. Through the analysis of subjective responses of participants, the authors demonstrate that the social pressure and shared indicators of availability (such as "the last time the user was online") provided by communication applications make people respond to personal messenger notifications more quickly. Moreover, their findings suggest that people tend to feel connected on receiving personal message notifications from their social ties. However, increase in the amount of such notifications could lead to increased stress levels of the recipients. Additionally, the authors found that the information triggered by proactive services are considered as least interesting by the users.

In [77], Mehrotra et al. investigated the role of fine-grained features of a delivered notification and an ongoing task on users' attentiveness and receptivity. They demonstrated that ringer mode, presentation of a notification, and engagement with a task all influence users' attentiveness. On the other hand, users' receptivity to notifications is influences by the type of information contained in the notification and the relationship between the sender and the recipient. Their findings also suggest that disruptive notifications are accepted by users only when they contain useful information. Finally, they demonstrated that the perceived disruption and response time to a notification both rely on users' personality.

In a different study, Mehrotra et al. conducted an in-depth analysis of interaction behavior with notifications in relation to the location and activity of users [74]. Their results demonstrated that users' attention toward new notifications (i.e., attentiveness) and willingness to accept them (i.e., receptivity) are strongly linked to the location they are in and also to their current activity. More specifically, they found that people are more receptive to notifications when they are at college, in libraries, on streets or residential areas, and least attentive to notifications at religious institutions. Moreover, their results demonstrated that people are least attentive to notifications while they are preparing to go to bed (or using the phone in bed) and during the time they exercise. People are found to be more receptive to notifications while they are exercising and doing chores (i.e., routine tasks).

In another study [78], Mehrotra et al. investigated the causal links between users emotional states and their interaction with mobile notifications. Their results demonstrated that people's activeness level has a significant association with their attentiveness and receptivity to mobile notifications. They also showed that people's attentiveness to notifications increases in stressful situations, which results in quicker responses to notifications arriving in that period. These results are inline with the findings of another study by Pejovic et al. to show that users' interruptibility for mobile notifications is significantly influenced by the level of users' engagement with their ongoing task [84].

In [29], Fischer et al. investigated the behavior of users on receiving notifications from specific categories of applications. They found that the reaction to notifications is a function of the importance of the corresponding application. In another study [30], Fischer et al. demonstrated that the importance of a notification depends on how interesting, actionable and relevant the delivered information for the recipient is. Findings of another study [28] suggest that applications should not trigger uninteresting and irrelevant notifications as people get annoyed and consider deleting applications that consistently triggers such notifications.

6.2 DESIGNING INTERRUPTIBILITY MANAGEMENT SYSTEMS FOR DELIVERING THE RIGHT INFORMATION AT THE RIGHT TIME

In this section we present different approaches taken by researchers and practitioners to design solutions for interruptibility management systems by focusing on delivering the right information at the right time. We have categorized these approaches based on the type of information used to build prediction models for managing interruptibility.

6.2.1   INTERRUPTIBILITY MANAGEMENT BY USING CURRENT ACTIVITY

In [45], the authors proposed the first interruptibility management tool for predicting the cost of interruptions triggered by phone calls. The system consisted of two Bayesian network models for predicting whether users will attend meetings on their calendar and the cost of being interrupted by incoming calls should a meeting be attended. More specifically, the first model was constructed to predict the interruptibility of a user, and the other model was constructed to predict the probability that users will attend meetings that appear on their electronic calendar. Finally, predictions from these two models were used to predict the expected cost of interruption at different times for a user. The models were trained by using the stream of sensed data generated by users' interaction with their computers and properties of items on their electronic calendar. Their results show that the models were able to predict interruptibility and attendance to meetings with the accuracy of 81% and 92%, respectively. It is worth remarking that this system was designed solely for managing phone calls, which were the only interruptions generated by mobile devices in their infancy when they were not equipped with a plethora of proactive mobile applications triggering numerous notifications everyday.

In [96], the authors discussed a personalized approach for predicting the cost of a mobile interruption by exploiting people's current activity data. They conducted a survey to understand mobile phone users' preferences and interruption cost in different situations. Their results suggest that the cost of interruptions vary across users. For instance, a user might not have any problem in receiving interruptions at work, while another user might consider these as a significant disruption. Therefore, according to their findings, interruptibility management models should be personalized.

6.2.2   INTERRUPTIBILITY MANAGEMENT BY USING THE TRANSITION BETWEEN ACTIVITIES

In [40], the authors explored the use of transitions between physical activities for delivering mobile notifications. More specifically, they conducted an experiment to compare users' receptivity to mobile notifications triggered at times corresponding to activity transition and at other times. Their study was based on the hypothesis that this transition indicates "self interruption" as users switch to another activity after its completion and the resistance to interruptions might be lower during such moments. The authors customized a few PDAs by adding two wireless accelerometers in each in order to capture users' physical movement. After using temporal smoothing on activity data, they captured four types of transitions: sitting to standing, standing to sitting, sitting to walking and walking to sitting. In order to learn users' interruptibility, users were triggered a questionnaire once every 10–20 min throughout the day, either randomly or at an activity transition. The questionnaire asked the users about their receptivity to notifications on a scale of 1 (very low) to 5 (very high). Their results show that interruptions delivered at a time corresponding to an activity switch are judged more positively compared to interruptions delivered at random times. These results suggest that the perceived disruption from mobile notifications could be significantly minimized by simply time-shifting the messages to moments when the user is transitioning between different physical activities.

In another study [29], Fischer et al. investigated the use of naturally occurring breakpoints during users' interaction with mobile phones as opportune moments to deliver mobile notifications. Participants were asked to report what were they doing on their phones at the time of notification arrival by means of an ESM questionnaire. Notifications for these questionnaires were triggered (i) at random times or (ii) after the user has finished a call or finished sending/reading an SMS. Through the analysis of around 2000 notifications collected from 20 participants over a period of 2 weeks, they found that notifications, which were delivered just after the user finished a call or sent/read a text message, have received quicker response. In other words, their findings suggest that users tend to handle notifications significantly more quickly after they finish an episode of mobile interaction than at random other times.

In [83], Okoshi et al. studied the use of breakpoints within users' interaction with a mobile phone for delivering notifications in order to reduce interruptions and improve users' experience. The authors developed Attelia—a system for detecting breakpoints in users' interaction with mobile phones and to defer notifications until such a breakpoint occurs. Attelia detects breakpoints during the user's interaction with a mobile phone in real-time, by using the sensors embedded in the phone. Attelia monitors users' interaction with applications and exploit this information in order to detect breakpoints. They used the NASA-TLX questionnaire [39] to quantify participants' subjective cognitive load. Based on a controlled study with 37 participants, authors demonstrated that the cognitive load of users (who were more sensitivity to interruptions) is reduced by 46% by triggering notifications at breakpoints compared to triggering notifications at random times. Later, they conducted an "in-the-wild" study involving 30 participants in order to validate their mechanism in a real-world scenario. The results of this "in-the-wild" study suggest that Attelia could reduce 33% of the cognitive load by delivering notifications at detected breakpoints. Moreover, the notifications delivered at breakpoints received a quicker response from users.

6.2.3   INTERRUPTIBILITY MANAGEMENT BY USING CONTEXTUAL DATA

Advances in the sensing capabilities of mobile phones have allowed for the monitoring of various context modalities, such as location, physical activity and colocation with other Bluetooth devices (i.e., colocation with other users). Numerous studies have demonstrated the potential of exploiting mobile sensing in order to infer not only numerous aspects of users' physical behavioral patterns [58, 76] but also their health and emotional states [14, 57, 62, 63, 66, 71, 92].

Scientists have used various aspects of the physical context of users, captured via mobile sensors, in order to construct machine learning-based models for predicting interruptibility of users. For example, Pejovic et al. developed InterruptMe [85]—an interruption management library for Android-based mobile devices. InterruptMe uses a mixed method of automated smartphone sensing to collect contextual information and experience sampling to ask users about their interruptibility at different moments. This information is exploited by InterruptMe to construct intelligent interruption models based on a series of machine learning algorithms for interruptibility prediction. To evaluate InterruptMe, the authors conducted a 2-week study with 20 participants and gathered users' *in-the-wild* contextual information such as location, physical activity, emotional state and the level of engagement with an ongoing task. Their results show that opportune moments for interruptions can be predicted with an average accuracy of 60% and the reported sentiments toward notification can also be predicted with a precision of 64% and a recall of 41%. Moreover, they found that the time taken by users to respond to notifications can be predicted accurately. Finally, they demonstrated that the online learning approach can be used to train models well enough to start making stable predictions within a week.

In [91], the authors conducted a survey with 84 users to understand people's reaction toward explicit indicators of their availability shared with others on messaging applications (such as "the last time the user was online"). Their results show that these indicators create a social pressure on users to respond to the messages but people still see a great value in sharing their attentiveness. However, the authors argued that the shared indicators of availability by messaging applications are weak predictors of recipients' attentiveness. Instead, machine-computed prediction of recipients' attentiveness should be used as a more reliable source. In order to validate their proposed approach, the authors conducted an experiment with 24 participants for a period of two weeks. They developed a mobile application that logs information about users' context and their actual attentiveness, including application name and the arrival time of notifications, response time (i.e., time taken to handle a notification from its arrival), launching and closing times of messaging applications, phone lock/unlock times, and the ringer mode. Using this data they computed 17 features and ranked them based on their entropy. They demonstrated that by using only the top seven features, a machine learning algorithm can construct a model that can predict users' attentiveness with an accuracy of 71%.

In [26], Dingler and Pielot argued that bounded-deferral strategies (i.e., strategies for deferring notifications up to a certain time period in order to reduce disruption caused by it) do not work if users are busy for long time periods. They suggested that a notification might lose its value if it is delayed for too long. Therefore, notifications must only be deferred if the phases of inattentiveness are brief. Based on their hypothesis they conducted a study to investigate whether users' attentiveness to mobile phone notifications can be predicted by using contextual information. Through a passive and continuous sensing approach, they collected phone-usage data from 42 participants for a period of 2 weeks. They demonstrated that users' attentiveness can be predicted using mobile phone usage with an accuracy of 80%. Their findings show that users are attentive to mobile notifications for around 12 h in a day. Also, the periods of users' inattentiveness to mobile notifications are often very short (i.e., 2–5 min).

Another study [88] explored the use of phone usage activity and contextual information for predicting users' attentiveness to calls. In order to collect data, the authors developed an application that temporarily mutes the ringer by simply shaking the phone. They logged anonymous data from 418 users corresponding to more than 31,000 calls mapped with recipients' context at the time of call arrival. They collected data which is available through the open API calls of the Android platform, such as physical activity, ringer mode, device posture, and time since last call. Their results demonstrated that by exploiting these features users' availability for calls can be predicted with 83% accuracy. Moreover, they showed that a personalized model training approach can increase the average accuracy by 87%. By using only the top five features (including last time the ringer mode was changed, last time the screen was locked/unlocked, current status of screen lock, last time the phone was plugged/unplugged from charging, time since last call) call availability can be predicted with an accuracy of 79%.

In [37], the authors argued that interruptibility prediction models often fail to infer the opportune moments to deliver information because they do not consider exploiting the sender-recipient relationship and the type of information. They proposed a theoretical framework to provide a different perspective on the design of interruption management systems by considering who the interruption is from or what it is about. In order to validate this, Mehrotra et al. [75] conducted a real-world study for over 3 weeks and collected around 70,000 instances of notifications from 35 users. They collected both the contextual data (such as location, activity, surrounding sounds and so on) and logs of users' interaction with mobile notifications. They grouped notifications according to the applications that initiated them, and the social relationship between the sender and the receiver. Consequently, they exploited users' contextual data, delivered information's type and the sender-recipient relationship for designing prediction models to learn the most opportune moment for the delivery of a notification carrying a specific type of information. Their results show that the notification receptivity can be predicted with sensitivity and specificity of 70% and 80%, respectively, which can go 10% higher for some users. Moreover, the authors demonstrate that their interruptibility prediction model outperforms an alternative approach based on user-defined rules of their own interruptibility (i.e., subjective rules with which users describe their interruptibility).

6.2.4   INTERRUPTIBILITY MANAGEMENT BY FILTERING IRRELEVANT INFORMATION

Previous studies have found that in order to not miss any newly available important information, people are willing to tolerate mobile notifications to interrupt them [52]. However, various mobile applications exploit their willingness by triggering a large number of notifications [90]. At the same time, some studies have demonstrated that users do not accept all notifications as their receptivity relies on the type and sender of information being delivered [75, 77]. For this reason, notifications that are uninteresting and irrelevant are mostly dismissed (i.e., swiped away without clicking) by users, which was also found in [30, 98]. Some examples include new game invites, app updates, predictive suggestions by recommendation system, and marketing emails. At the same time, previous studies have shown continual trigger of such irrrelevant notifications becomes a cause of annoyance for users, which could result in uninstalling the corresponding application [28, 98]. These findings suggest that interruptibility management system should take into consideration the relevance and users' interest toward delivered notifications.

In [30] Fisher et al. examined whether the user's receptivity to a notification is influenced by its content and the time of delivery. In order to understand the role of notification content, they recruited 11 participants and asked them to report their interest for the given 28 categories of content on a 7-point Likert scale. For each participant, the content types rated (by the same participant) as top 3 were considered as "good content" and the lowest three were considered as "bad content." Moreover, participants were asked to specify the time window during which they were interested in receiving notifications of these types. Participants received six notifications (three each of good and bad content) every day. These notifications were delivered at three opportune times and three inopportune ones. Their results show that users' receptivity is significantly higher for good content compared to bad content. However, there was no significant differences in users' receptivity at opportune and inopportune times. This might suggest that users' receptivity is associated with notification content rather than the time of delivery. Furthermore, they collected the subjective responses of participants to investigate the role of notification content's characteristics for influencing users' receptivity. Their findings suggest that users' interest, entertainment, relevance and the actions required to handle the notification significantly influence their receptivity.

In a recent study [70, 72], the authors have presented the design of a solution for managing interruptibility by discovering rules for users' receptivity in different contexts through the analysis of their past interaction with notifications. They first evaluate their approach on the My Phone and Me study's dataset [77] that comprises users' interaction with "in-the-wild" notifications and context data. More specifically, they grouped notifications by exploiting their titles and, then, constructed the association rules by using the combinations of notification groups and context modalities including activity, time and location. Through an extensive they have demonstrated that, by exploiting only the user's location and the title of a notification, their system could predict with 91% precision if that notification will be accepted or declined.

Furthermore, they funneled their findings into the development of *PrefMiner*[1]—an Android library for notification management that offers an API to extract rules for the user's receptivity. Interestingly, unlike previous interruptibility studies, Mehrotra et al. not just performed the offline evaluation of their mechanism on the collected data, but they conducted another experiment to carry out an in-the-wild study of the usage of PrefMiner. In order to do this, they exploited the discovered association rules by making them transparent to users so that they can

---

[1]Available at: https://github.com/AbhinavMehrotra/PrefMiner

check their appropriateness. The results of their in-the-wild study demonstrated that PrefMiner suggested 179 rules out of which 57% were accepted by users. These rules were able filter out unwanted notifications with 46% accuracy.

6.3   DESIGNING INTERRUPTIBILITY MANAGEMENT SYSTEMS FOR MULTI-DEVICE ENVIRONMENTS

Smart devices are becoming more and more ubiquitous and have arrived in our everyday lives. In today's ubiquitous computing environment users carry and interact with an increasing number of smart mobile devices such as laptops, smartphones, tablets, and smartwatches. One of the core features of smart devices is the ability to run different versions of the same application and, thus, allowing users to interact with these apps at anytime from any device. Such applications, which are supported on multiple devices, are called cross-platform applications. These applications leverage notifications to trigger real-time alerts for steering users' attention toward newly available information through auditive, visual and haptic signals.

Indeed, these notifications are beneficial to users for proactive personalized information delivery about a variety of events. However, as discussed in previous chapters, notifications sometimes arrive at inappropriate moments and, for this reason, they can have an adverse impact on the execution of the ongoing tasks [8, 21, 22] and even on the affective state of users [2, 7]. The problem is exacerbated in the multi-device environment due to the fact that cross-platform applications prompt users on multiple devices at the same time, which could make these notifications even more disruptive and annoying.

As discussed in the previous section, until now, previous interruptibility management studies have focused on inferring opportune moments for delivering mobile notifications (e.g., [40, 64, 75, 93, 94, 101]) and learning the types of notifications users prefer to receive in different situations (e.g., [70, 89]). In the recent years, since the smart devices have arrived in our everyday lives, the problem of delivering notifications *on the right device* in different contexts has become the focus of the research community. In this section we present different approaches taken by researchers and practitioners for designing solutions for managing interruptibility by delivering the information through the right medium.

6.3.1   UNDERSTANDING USERS' PERCEPTION TOWARD INTERRUPTIONS CAUSED BY NOTIFICATIONS IN A MULTI-DEVICE ENVIRONMENT

Since previous studies on interruptibility management have focused on notifications delivered on a single device. There was a lack of knowledge for understanding how future notification systems should be designed by considering multi-device environment. In order to bridge this gap, in [104] the authors investigated how users' perceive notifications in multi-device environments.

They conducted a study with 16 participants who used 4 different types of smart devices (i.e., smartphone, tablet, smartwatch, and PC), which are the most commonly used smart mobile devices capable of displaying notifications. Over the course of one week they collected logs of users' interaction with all four devices, such as display on/off, connection status (WiFi, mobile data, or offline), power on/off, headset connected, and charging connected or not. Also, they logged users' current location and the current activity. In addition, they asked users to respond to questionnaires, triggered every 45–90 min, comprising of the following four multiple choice questions.

- Where are you?

- How many people are in your surroundings?

- The mentioned device is in my proximity.

- I want to receive a notification on the mentioned device.

The analysis of their data revealed that participants preferred to be notified on the smartphone, followed by the smartwatch, the PC, and the tablet. Their results demonstrated that users' receptivity to notifications on a specific device is influenced by their proximity to that device. They also showed that the participants prefer to receive notifications on devices with which they are currently interacting. Moreover, devices that are still for a longer period indicates that they are less suitable for delivering notifications. Participants also indicated that they prefer to receive notifications on smart watches when they are at work place. Overall, their results show that users prefer to receive notifications on specific devices based on their situation.

6.3.2   DESIGNING INTERRUPTIBILITY MANAGEMENT SYSTEMS FOR DELIVERING INFORMATION THROUGH THE RIGHT MEDIUM

In [73], the authors conducted the first study to address the problem of delivering notifications on the right device in a multi-device environment. The authors suggest that since it is very difficult to define the concept of right device, but as a first approximation, it can be seen as the device on which users prefer to handle a specific notification given their context.

In order to design a solution for intelligent cross-platform notification delivery, they first conducted an in-situ study to investigate the contextual factors impacting users' decisions for handling notifications on specific devices. In particular, they investigated how users behave in different contexts when they receive a notification on their mobile phone as well as on a generic alternative device (i.e., any alternative computing device that runs a version of applications that are also available for mobile phones) at the same time.

Through a passive logging smartphone application they collected around 25,000 in-the-wild notifications triggered by cross-platform applications on smart devices of 24 users over a period of 21 days. Moreover, their app also passively collected information about users' interaction with phone (including screen on/off and touch events) and context information (including physical activity, location and network connectivity). In order to infer whether a notification is handled or not (i.e., handled on some other device), they assume that a notification is automatically removed from the notification bar (or from the lock screen) of the phone if it was delivered on some other device and the user has already interacted with it on other device. More specifically, their assumption is that a notification is handled on a mobile phone only if the user has interacted with the phone between its arrival time (t_a) and its removal time (t_r). If there was no phone interaction logged in the interval between the time t_a and t_r, the notification is assumed to be handled on an alternative device.

In their dataset, they found that around 29% of notifications (triggered by cross-platform applications) were handled on an alternative device (i.e., not handled on the mobile phone). They first investigated the effect of different factors on users' behavior with respect to notification handling in a multi-device environment. Their results demonstrated that there is a significant relationship between the user's behavior in terms of handling notifications and several factors, including activity, location, network connectivity, application category, and the device used for attending to the previous notification. However, they did not find any significant impact of the arrival time of notifications on the user's behavior for handling them.

Based on these findings, the authors designed and implemented—NotifyMeHere—a solution for intelligent notification delivery in multi-device environments [73]. Their systems used information about users' activity, location, network connectivity, application category, and the previous click feature (i.e., the device used for attending to the previous notification) in order to predict the device on which a notification will be handled by a user. They found that the previous click feature could solely be used for predicting where the next notification would be handled. However, with further in-depth analysis they demonstrated that the previous click feature can be used alone for predicting the device for delivering notifications but only when the preceding notification has arrived approximately within 10 min. Once the threshold on 10 min is passed, other features become more dominant for predicting the user's notification handling behavior. Finally, they constructed and evaluated a set of prediction models, considering both generalized and personalized training approaches. Their results have shown that an individualized model is characterized by better prediction performance (i.e., 82% specificity and 91% sensitivity) compared to generalized model's performance (68% specificity and 90% sensitivity). Furthermore, in their study, the authors have also demonstrated the implementation of an online predictor that achieved a specificity of 70% and a sensitivity of 95% in just 7 days.

6.4   INTERRUPTIBILITY MANAGEMENT—AN ALTERNATIVE APPROACH FOR CONTEXT-AWARE ASSISTIVE APPS

Today numerous context-aware apps facilitate spatial- and temporal-aware features. For instance, apps for location based information alert, and reminder for a scheduled event. However, these apps do not consider the amount of distraction caused by their service, which is directly dependent on how long the pushed information interrupts a primary activity. In other words, the cost of interruption is based on the amount of attention required to handle the pushed information.

In [99], the authors proposed to categorize activities (for which the distraction is introduced) in four categories—Snap, Pause, Tangent, and Extended. This categorization can be done based on the amount of time required out of the user's primary activity to handle the introduced distraction. The authors suggested to design a *distraction matrix* to categorize each activity as Snap, Pause, Tangent, or Extended.

Activities that are usually completed within a few seconds are categorized as "Snap." For instance, a reminder about the upcoming event, where the user does not have to interrupt their primary task. The activities belonging "Pause" category require users attention for a few minutes, which causes a task switch from the primary activity to the new activity, but the user could return back to the primary activity within a few minutes. For instance, attending a phone call. The "Tangent" activities are medium length tasks and usually irrelevant to the user's primary task. For example, a message from a colleague for a quick discussion on another project. Activities belonging to the "Extended" category introduce a deliberate switch to another task, and there is a significant delay to return back to the primary task. For example, a message from a friend to go for lunch.

Anhalt et al. suggested that the goal of context-aware assistive apps should be to move distraction activities toward the left side of the matrix (i.e., as Snap activities that take less time from the user's primary activity) [99]. In this way, such apps could benefit users by minimizing the overall distractions and, thus, realizing the vision of calm computing.

6.5   SUMMARY

There have been numerous approaches proposed for designing an interruptibility management system for mobile devices. In Tables 6.1, 6.2, and 6.3 we present the summary of the literature in this area. Studies focused on designing interruptibility management system for mobile devices have considered two key dimensions: (i) delivering right information at the right time and (ii) delivering information through the right medium. More specifically, *right time* indicates that the information should be delivered when the user is likely to quickly attend it, whereas, *right information* refers to delivering the information that is considered relevant to the user in the current context. On the other hand, the *right medium* indicates the device on which the information should be delivered to quickly gain users' attention without causing much disruption.

There have been several studies focusing on managing interruptibility by predicting opportune moments for interrupting users for delivering the right information at the right time. These studies have investigated how users perceive interruptions from mobile notifications. For example, it has been established that users receive a plethora of notification every day and they try to manage interruptions themselves by configuring the ringer mode or stopping notifications from specific apps. Sometimes users uninstall an app if it is triggering too many notifications at

Table 6.1: Key findings of the studies focusing on understanding users' perception toward interruptibility in mobile environment

Key Findings	Study
People receive around 60-100 notifications each day	[75, 98]
Reported as disruptive but useful for information awareness	[52]
People handle notifications within a few minutes from their arrival	[75, 98]
People manage interruptibility through the ringer mode configurations	[16]
People stop notifications from a specific applications	[65, 108]
Varying importance is given by people to notifications triggered by different applications	[98]
People have social pressure to quickly respond to notifications	[90]
People's receptivity to notifications is influenced by their context, sender-recipient relationship, and notification content	[74, 77]
People's receptivity to notifications is influenced by their mood	[78]
People tend to delete apps that keep sending notifications at inopportune moments	[28]

inopportune moments. There is a social pressure on users to respond to notifications, but their context and mood also influence their interruptibility.

On the other hand, predicting the right medium to deliver notification for managing interruptibility has instead become the focus of the research community in the recent years. Users' receptivity to notifications on a specific device is influenced by their proximity to that device. They prefer to receive notifications on devices with which they are currently interacting, and on smart watches when they are at work place. Moreover, users' behavior in terms of handling notifications in multi-device environment is influenced by their activity, location, network connectivity, application category and the device used for attending to the previous notification.

We believe that the studies presented in this chapter have useful implications for designing interruptibility management systems for existing and future devices. In the next chapter, we will present some open challenges that remain to be addressed for realizing an ideal solution for interruptibility management systems.

Table 6.2: Studies in the area of understanding and modeling interruptibility in mobile environment

Types of Information Used for Interuptibility Prediction	Types of Approach
Interruptibility Management by Using Current Activity	Predicting the level of interruption through calls by using users' activity and calendar properties [45]
	Predicting and automatic setting of the phone ringer by exploiting the activity [96]
Interruptibility Management by Using the Transition between Activities	Using transitions between physical activities for delivering mobile notifications [40]
	Using breakpoints during users' interaction with mobile phones for delivering notifications [29, 83]
Interruptibility Management by Using Contextual Data	Predicting receptivity by using context information [85]
	Predicting attentiveness to notication by using mobile phone usage features [26, 91]
	Predicting call availability by using the phone usage activity and contextual information [88]
	Predicting receptivity by using the notification content (i.e., information type and relationship with the sender) and context information [75]
Filtering Irrelevant Information	Notifications that are uninteresting or irrelevant to people's interests are mostly dismissed [98]
	Predicting receptivity to notifications by the predefined relevant and irrelevant informations classes [30]
	Predicting users' preference about the types of information they want to receive in specific contexts [70, 72]

Table 6.3: Key findings of the studies focusing on designing solutions for managing interruptibility by delivering the information through the right medium

Key Findings	Study
People prefer to be notified on the smartphone, followed by the smartwatch, PC, and tablet	[104]
People's receptivity to notifications on a specific device is influenced by their proximity to that device.	
People prefer to receive notifications on devices with which they are currently interacting.	
People prefer to receive notifications on smartwatches when they are at work place.	
People handle around 71% of overall notifications (triggered by cross-platform applications) on smartphones	[73]
People's behavior in terms of handling notifications is influenced by their activity, location, network connectivity, application category, and the device used for attending to the previous notication	
Proposed a solution for intelligent notification delivery in multi-device environments that achieves 82% specificity and 91% sensitivity	

CHAPTER 7

Limitations of the State of the Art and Open Challenges

Interruptions are an inevitable part of our daily life. As discussed in this book, the effects of disruption caused by interruptions occurring at inopportune moments have been studied thoroughly in the past. Numerous studies have been conducted to investigate the effect of interruptions on users' ongoing tasks. More specifically, researchers have found that completion time [21, 22, 81], error rate [60], and even emotions toward the ongoing tasks [2, 7] are adversely affected by interrupting users at inappropriate moments. However, studies have also provided evidence that in order to not miss any newly available important information, people tolerate some disruption [52].

Since the era of desktops, managing interruptions has been a key theme in Human-Computer Interaction research [3, 43, 44, 49]. With the advent of mobile and wearable technologies, the problem of managing interruptibility has become even more pressing as users can now receive notifications anywhere and at anytime. Indeed, numerous research efforts have been carried out with the goal of designing interruptibility management system for mobile environments [40, 52, 70, 75, 83, 85]. Most of the work on mobile interruptibility emphasizes the exploitation of features that can easily be captured through mobile sensors, such as task phases [29, 40, 52], users' context including location and activity [26, 85, 91], and notification content [70, 75] to infer the right time to interrupt. More specifically, studies have shown that notifications are considered more positively and received a faster response when delivered while a user switches from one activity to another [29, 40]. On the other hand, studies have also demonstrated that machine learning algorithms can learn about users' interruptibility by exploiting passively sensed contextual information and notification content [75, 85]. Also, studies have demonstrated that the contextual information can be exploited to infer and filter out the irrelevant information from being delivered [70]. Furthermore, a handful of recent studies have investigated the feasibility of predicting the right medium to deliver notification for managing interruptibility [73, 104]. The studies have shown that users' behavior in terms of handling notifications in multi-device environment can be modeled by exploiting the contextual and phone interaction data [73].

Consequently, existing studies have focused on various challenges concerning the understanding and learning users' behavior in terms of interactions with notifications. However, the characterization of attentiveness and receptivity of users for mobile notifications is still an open problem. We believe that there is still a considerable scope for improvement, for example by exploiting other physical, social and cognitive factors for modeling users' notification interaction behavior. For instance, more knowledge about users' cognitive context could help the system to reduce the amount of notifications delivered to users when they are stressed [74]. We now summarize some key open questions in the area that must be investigated to build intelligent mechanisms that could effectively trigger the *right* information in a given context.

7.1 DEFERRING NOTIFICATIONS

Let us start from a key open question in this area: should we defer a notification if it is not delivered at an opportune moment and for how long. Until now, interruptibility management studies have focused on inferring if the current moment is opportune to deliver notifications or not. We believe that in order to be more effective, these systems should not just predict users' current interruptibility, but if the current time is not an opportune one, it should also *anticipate* the best moment in the nearest future [79]. This would enable the overlying application to decide whether the notification should be deferred until the predicted opportune moment or not.

A recent study has found that users tend to defer notifications related to people and events [103]. They also found that user decision for deferring notifications are influenced by their daily routines. We believe that such findings and prediction techniques similar to the ones employed for other interruptibility management systems could be exploited in order to address this challenge.

7.2 MONITORING COGNITIVE CONTEXT

Previous approaches for interruptibility management focus on exploiting users' physical context. However, as discussed in this lecture, users' interruptibility might also be associated with their cognitive context. For instance, previous studies have already have already reported that users' interruptibility is influenced by cognitive factors such as their mood [78], their engagement with the current task [84], and complexity of the interrupting task [77]. However, none of the studies have exploited this information to build an interruptibility prediction model.

We believe that this limitation exists due to the fact that current mobile and wearable technologies are yet not capable of passively sensing such information. So, there is indeed a need for developing and evaluating mechanisms for automatically capturing the level of users' engagement with the current task, complexity and difficult of execution of the interrupting task, and similar cognitive factors that might influence interruptibility. One of the potential approaches might be to explore the use of affective computing [87] in order to monitor users' emotional states.

7.3   LEARNING "GOOD" BEHAVIOR: INTERRUPTIONS FOR POSITIVE BEHAVIOR INTERVENTION

Until now, all interruptibility studies have focused on learning the observed user behavior associated with the sensed contextual information. However, interruptibility management system can also be considered as a key component for behavior change intervention tools that could help prevent and modify *harmful behavior* of users [59]. In other words, a potential future direction of these systems could also be to gather the knowledge about *good behavior* and exploit it to improve the behavior of users. Such knowledge about *good behavior* for interacting with notifications could potentially be obtained by carrying out a large-scale ESM-based study that can query users about the ideal notification-interaction behavior in their current situation. However, there is an inherent problem related to learning "good" vs. "bad" behavior. The problem is inherent in the fact that a machine learning might not distinguish between a behavior that should be promoted and one that should not. As an example, let us consider a learning component that is able to learn the right moment to interrupt by past experience. If upon receiving a notification a user reads emails on their mobile phone while driving, the notification mechanism should *not* learn this behavior and deliver emails accordingly. Instead, the mechanism should infer that it is a *harmful behavior* to read notifications while driving and try to avoid sending unnecessary emails. Indeed, if the information is critical it should be delivered immediately regardless of the current situation. However, designing such an ideal notification delivery mechanism is extremely difficult and have never been considered in the scope of any interruptibility study.

7.4   MODELING FOR MULTIPLE DEVICES

A previous study has shown that users' prefer to receive notifications on specific devices based on their situation [104]. Inline to this work, another study has proposed a solution for modeling users' behavior in terms of handling notifications in multi-device environment by exploiting the contextual and phone interaction data in order to deliver notifications on the right device. However, they conducted a pilot study by considering only two devices: mobile phones and alternatives (i.e., any device other than phones).

Consequently, the aspect of delivering information through the right medium is not well studied. Almost all of the previous studies have not focused on predicting users' behavior on receiving cross-platform notifications, which are delivered on multiple devices at the same time. In other words, there is a lack of understanding of the features that determine users' receptivity to such notifications on a specific device in a given context. Given the fact that users are surrounded by an increasing number of devices that are able to receive a notifications (such as apps in laptops, mobile phones, wearables, smart television sets, and appliances), the design of such mechanism is an open and interesting research area.

7.5   NEED FOR LARGE-SCALE STUDIES

Almost all studies in the area of interruptibility management are conducted with small samples of the population and for short time periods. At the same time, these studies are often publicized through the network of the researchers performing the studies. This could introduce a bias deriving from the self-selected sample of users and thus the behavior of a certain group of user (within a network) might be different from others. Moreover, the validation of the interruptibility prediction mechanisms in these studies is usually performed in an offline fashion (i.e., the evaluation is performed on the collected data *a posteriori*). For this reason, the results presented in these studies do not have ecological validity as the collected datasets might be biased toward a certain group of population.

Therefore, we believe that there is a need for large scale studies as well as *in-the-wild* deployments [19, 95] to guarantee the ecological validity and robustness of the proposed interruptibility prediction mechanisms. We also believe that reproducing these studies in different social context and users' demographics is also essential.

We believe that the open challenges presented in this chapter have useful implications for designing the future interruptibility management systems. In the next chapter, we will summarize the content of this lecture and hope they are useful for realizing an ideal solution of the interruptibility management system.

CHAPTER 8

Summary

The always-on connectivity of smart mobile devices (such as phones, watches, and tablets , etc.) have made them a unique platform to push information in real time and, thus, they represent a medium for receiving information in an effortless way. However, due to the mobile nature of these devices the inevitable notifications often arrive at inopportune moments. In this book we discussed the key studies in the area of interruptibility management, which demonstrated that such disruptive notifications can adversely affect the ongoing task and affective state of the user. Moreover, with the advent of advanced technologies in mobile phones, this tension is exacerbated as individuals have to deal with a plethora of mobile notifications everyday, some of which are disruptive, and on multiple devices at the same time. Overall, these findings provide evidence for the need of a smart notification mechanism that can reduce the level of disruption by filtering out the irrelevant notifications and delivering the information at opportune moments through the right medium.

We have also devoted special attention to the discussion of the state-of-the-art in modeling and anticipating interruptible moments. Desktop interruptibility management mechanisms have mainly focused on task transition phases, where tasks are related to interaction with the device itself. On the other hand, in existing mobile interruptibility studies, anticipatory models predicts users' interruptibility by relying on their physical contextual information sampled through mobile sensors and the content of notifications. Furthermore, we note that apart from handful of interruptibility management mechanisms most models are evaluated only in the offline settings. This is due to the fact that performance of these intelligent mechanisms do not always meet users' expectations in terms of usability and correctness. Indeed, learning the behavior of users for interacting with notifications is not an easy task as it relies on various contextual modalities, both physical and cognitive. Furthermore, we also discussed some recent studies focused on intelligent notification delivery in multi-device environments in order to reduce interruptions by predicting the right medium to deliver notification.

Moreover, we note that all the proposed interruptibility management mechanisms aim at learning the observed user behavior associated with the sensed contextual information and adapting the notification delivery process accordingly. Ideally, such mechanisms should also possess the knowledge about *good behavior* that can be exploited to improve the behavior of users. More in general, intelligent notification systems can be used to deliver intelligent mobile systems.

We believe that further improvements in smart devices' contextual inference capabilities would also enable researchers to enhance our understanding of users' interaction with notifications on these devices. We hope that the critical discussion of interruptibility studies and the overview of the key open challenges presented in this book will be of valuable help for researchers and practitioners working in this exciting and important area.

Bibliography

[1] Aalto, L., Göthlin, N., Korhonen, J., and Ojala, T. (2004). Bluetooth and WAP push based location-aware mobile advertising system. In *MobiSys'04*, pp. 49–58, ACM. DOI: 10.1145/990064.990073 1

[2] Adamczyk, P. D. and Bailey, B. P. (2004). If not now, when?: The effects of interruption at different moments within task execution. In *CHI'04*, pp. 271–278, ACM. DOI: 10.1145/985692.985727 2, 12, 14, 22, 37, 45

[3] Adamczyk, P. D., Iqbal, S. T., and Bailey, B. P. (2005). A method, system, and tools for intelligent interruption management. In *Workshop on Task Models and Diagrams*, pp. 123–126, ACM. DOI: 10.1145/1122935.1122959 22, 45

[4] Atkinson, J. W. (1953). The achievement motive and recall of interrupted and completed tasks. *Journal of Experimental Psychology*, 46(6):381–390. DOI: 10.1037/h0057286 15

[5] Edwards, M. B. and Gronlund, S. D. (1998). Task interruption and its effects on memory. *Memory*, 6(6):665–687. DOI: 10.1080/741943375 11, 12

[6] Baddeley, A. D. (1976). *The Psychology of Memory*. Basic Books. DOI: 10.2307/1421745 11

[7] Bailey, B. P. and Konstan, J. A. (2006). On the need for attention-aware systems: Measuring effects of interruption on task performance, error rate, and affective state. *Computers in Human Behavior*, 22(4):685–708. DOI: 10.1016/j.chb.2005.12.009 2, 37, 45

[8] Bailey, B. P., Konstan, J. A., and Carlis, J. V. (2000). Measuring the effects of interruptions on task performance in the user interface. In *SMC'00*, pp. 757–762, IEEE. DOI: 10.1109/icsmc.2000.885940 2, 11, 12, 13, 37

[9] Bailey, B. P., Konstan, J. A., and Carlis, J. V. (2001). The effects of interruptions on task performance, annoyance, and anxiety in the user interface. In *INTERACT'01*, pp. 593–601, IEEE. 12, 13, 14

[10] Beatty, J. (1982). Task-evoked pupillary responses, processing load, and the structure of processing resources. *Psychological Bulletin*, 91(2):276–292. DOI: 10.1037//0033-2909.91.2.276 22, 26

[11] Begole, J. B., Matsakis, N. E., and Tang, J. C. (2004). Lilsys: Sensing unavailability. In *CSCW'04*, pp. 511–514, ACM. DOI: 10.1145/1031607.1031691 21, 25, 26

[12] Braune, R. and Wickens, C. D. (1986). Time-sharing revisited: Test of a componential model for the assessment of individual differences. *Ergonomics*, 29(11):1399–1414. DOI: 10.1080/00140138608967254 15

[13] Cabon, P., Coblentz, A., and Mollard, R. (1990). Interruption of a monotonous activity with complex tasks: Effects of individual differences. *Human Factors and Ergonomics Society Annual Meeting*, 34(13):912–916. DOI: 10.1177/154193129003401302 16

[14] Canzian, L. and Musolesi, M. (2015). Trajectories of depression: Unobtrusive monitoring of depressive states by means of smartphone mobility traces analysis. In *UbiComp'15*, pp. 1293–1304, ACM. DOI: 10.1145/2750858.2805845 33

[15] Card, S. K., Newell, A., and Moran, T. P. (1983). *The Psychology of Human–Computer Interaction*. L. Erlbaum Associates Inc. DOI: 10.1201/9780203736166 23

[16] Chang, Y.-J. and Tang, J. C. (2015). Investigating mobile users' ringer mode usage and attentiveness and responsiveness to communication. In *MobileHCI'15*, pp. 6–15, ACM. DOI: 10.1145/2785830.2785852 30, 41

[17] Clark, H. H. (1996). *Using Language*. Cambridge University Press. DOI: 10.1017/cbo9780511620539 7, 9

[18] Clark, H. H. and Schaefer, E. F. (1989). Contributing to discourse. *Cognitive Science*, 13(2):259–294. DOI: 10.1207/s15516709cog1302_7 7

[19] Consolvo, S., Bentley, F. R., Hekler, E. B., and Phatak, S. S. (2017). Mobile user research: A practical guide. *Synthesis Lectures on Mobile and Pervasive Computing*, 9(1):i–195. DOI: 10.2200/s00763ed1v01y201703mpc012 48

[20] Cortes, C. and Vapnik, V. (1995). Support-vector networks. *Machine Learning*, 20(3):273–297. DOI: 10.1007/bf00994018 24

[21] Cutrell, E., Czerwinski, M., and Horvitz, E. (2001). Notification, disruption, and memory: Effects of messaging interruptions on memory and performance. In *Interact'01*, pp. 263–269, IOS Press. 2, 11, 12, 13, 21, 22, 26, 37, 45

[22] Czerwinski, M., Cutrell, E., and Horvitz, E. (2000a). Instant messaging and interruption: Influence of task type on performance. In *OZCHI'00*, pp. 361–367, ACM. 2, 12, 13, 37, 45

[23] Czerwinski, M., Cutrell, E., and Horvitz, E. (2000b). Instant messaging: Effects of relevance and timing. *People and Computers XIV: Proceedings of HCI*, 2:71–76. 11, 12, 13, 21, 22, 26

[24] Dabbish, L., Mark, G., and González, V. M. (2011). Why do I keep interrupting myself?: Environment, habit and self-interruption. In *CHI'11*, pp. 3127–3130, ACM. DOI: 10.1145/1978942.1979405 6

[25] Dahlbäck, N., Jönsson, A., and Ahrenberg, L. (1993). Wizard of Oz studies—why and how. *Knowledge-Based Systems*, 6(4):258–266. DOI: 10.1016/0950-7051(93)90017-n 24

[26] Dingler, T. and Pielot, M. (2015). I'll be there for you: Quantifying attentiveness towards mobile messaging. In *MobileHCI'15*, pp. 1–5, ACM. DOI: 10.1145/2785830.2785840 34, 42, 45

[27] Dix, A., Finlay, J., Abowd, G., and Beale, R. (1993). *Human-Computer Interaction*. Prentice Hall. DOI: 10.1007/978-1-4614-8265-9_192 11

[28] Felt, A. P., Egelman, S., and Wagner, D. (2012). I've got 99 problems, but vibration ain't one: A survey of smartphone users' concerns. In *SPSM'12*, pp. 33–44, ACM. DOI: 10.1145/2381934.2381943 3, 31, 36, 41

[29] Fischer, J. E., Greenhalgh, C., and Benford, S. (2011). Investigating episodes of mobile phone activity as indicators of opportune moments to deliver notifications. In *MobileHCI'11*, pp. 181–190, ACM. DOI: 10.1145/2037373.2037402 31, 33, 42, 45

[30] Fischer, J. E., Yee, N., Bellotti, V., Good, N., Benford, S., and Greenhalgh, C. (2010). Effects of content and time of delivery on receptivity to mobile interruptions. In *MobileHCI'10*, pp. 103–112, ACM. DOI: 10.1145/1851600.1851620 2, 19, 31, 36, 42

[31] Fogarty, J., Hudson, S. E., Atkeson, C. G., Avrahami, D., Forlizzi, J., Kiesler, S., Lee, J. C., and Yang, J. (2005a). Predicting human interruptibility with sensors. *ACM Transactions on Computer-Human Interaction*, 12(1):119–146. DOI: 10.1145/1057237.1057243 24, 26

[32] Fogarty, J., Hudson, S. E., and Lai, J. (2004). Examining the robustness of sensor-based statistical models of human interruptibility. In *CHI'04*, pp. 207–214, ACM. DOI: 10.1145/985692.985719 22, 26

[33] Fogarty, J., Ko, A. J., Aung, H. H., Golden, E., Tang, K. P., and Hudson, S. E. (2005b). Examining task engagement in sensor-based statistical models of human interruptibility. In *CHI'05*, pp. 331–340, ACM. DOI: 10.1145/1054972.1055018 22, 24, 25, 26

[34] Foth, M. and Schroeter, R. (2010). Enhancing the experience of public transport users with urban screens and mobile applications. In *MindTrek'10*, pp. 33–40, ACM. DOI: 10.1145/1930488.1930496 1

[35] Friedman, N., Geiger, D., and Goldszmidt, M. (1997). Bayesian network classifiers. *Machine Learning*, 29(2-3):131–163. DOI: 10.1002/9780470400531.eorms0099 24

[36] Gillie, T. and Broadbent, D. (1989). What makes interruptions disruptive? A study of length, similarity, and complexity. *Psychological Research*, 50(4):243–250. DOI: 10.1007/bf00309260 11, 12, 13

[37] Grandhi, S. A. and Jones, Q. (2009). Conceptualizing interpersonal interruption management: A theoretical framework and research program. In *HICSS'09*, pp. 1–10, IEEE. DOI: 10.1109/hicss.2009.124 35

[38] Grosz, B. J. and Sidner, C. L. (1986). Attention, intentions, and the structure of discourse. *Computational Linguistics*, 12(3):175–204. 5

[39] Hart, S. G. and Staveland, L. E. (1988). Development of NASA-TLX (task load index): Results of empirical and theoretical research. *Advances in Psychology*, 52:139–183. DOI: 10.1016/s0166-4115(08)62386-9 33

[40] Ho, J. and Intille, S. S. (2005). Using context-aware computing to reduce the perceived burden of interruptions from mobile devices. In *CHI'05*, pp. 909–918, ACM. DOI: 10.1145/1054972.1055100 32, 37, 42, 45

[41] Hoeks, B. and Levelt, W. J. (1993). Pupillary dilation as a measure of attention: A quantitative system analysis. *Behavior Research Methods, Instruments, and Computers*, 25(1):16–26. DOI: 10.3758/bf03204445 22, 26

[42] Horvitz, E., Apacible, J., and Subramani, M. (2005a). Balancing awareness and interruption: Investigation of notification deferral policies. In *UM'05*, pp. 433–437, Springer. DOI: 10.1007/11527886_59 22, 26

[43] Horvitz, E., Jacobs, A., and Hovel, D. (1999). Attention-sensitive alerting. In *UAI'99*, pp. 305–313, Morgan Kaufmann Publishers Inc. 21, 23, 26, 45

[44] Horvitz, E., Koch, P., and Apacible, J. (2004). BusyBody: Creating and fielding personalized models of the cost interruption. In *CSCW'04*. DOI: 10.1145/1031607.1031690 21, 25, 26, 45

[45] Horvitz, E., Koch, P., Sarin, R., Apacible, J., and Subramani, M. (2005b). Bayesphone: Precomputation of context-sensitive policies for inquiry and action in mobile devices. In *User Modeling'05*, pp. 251–260, Springer. DOI: 10.1007/11527886_33 32, 42

[46] Hudson, S., Fogarty, J., Atkeson, C., Avrahami, D., Forlizzi, J., Kiesler, S., Lee, J., and Yang, J. (2003). Predicting human interruptibility with sensors: A Wizard of Oz feasibility study. In *CHI'03*, pp. 257–264, ACM. DOI: 10.1145/642611.642657 21, 24, 26

[47] Husain, M. (1987). Immediate and delayed recall of completed-interrupted tasks by high and low anxious subjects. *Manas*, 34(2):67–71. 16

[48] Iqbal, S. T., Adamczyk, P. D., Zheng, X. S., and Bailey, B. P. (2005). Towards an index of opportunity: Understanding changes in mental workload during task execution. In *CHI'05*, pp. 311–320, ACM. DOI: 10.1145/1054972.1055016 17, 22, 26

[49] Iqbal, S. T. and Bailey, B. P. (2006). Leveraging characteristics of task structure to predict the cost of interruption. In *CHI'06*, pp. 741–750, ACM. DOI: 10.1145/1124772.1124882 23, 26, 45

[50] Iqbal, S. T. and Bailey, B. P. (2007). Understanding and developing models for detecting and differentiating breakpoints during interactive tasks. In *CHI'07*, pp. 697–706, ACM. DOI: 10.1145/1240624.1240732 22, 23

[51] Iqbal, S. T. and Bailey, B. P. (2010). Oasis: A framework for linking notification delivery to the perceptual structure of goal-directed tasks. *ACM Transactions on Computer-Human Interaction*, 17(4):15. DOI: 10.1145/1879831.1879833 21, 25

[52] Iqbal, S. T. and Horvitz, E. (2010). Notifications and awareness: A field study of alert usage and preferences. In *CSCW'10*, pp. 27–30, ACM. DOI: 10.1145/1718918.1718926 2, 30, 35, 41, 45

[53] Iqbal, S. T., Zheng, X. S., and Bailey, B. P. (2004). Task-evoked pupillary response to mental workload in human-computer interaction. In *CHI'04*, pp. 1477–1480, ACM. DOI: 10.1145/985921.986094 21, 22, 26

[54] Joslyn, S. and Hunt, E. (1998). Evaluating individual differences in response to time-pressure situations. *Journal of Experimental Psychology: Applied*, 4(1):16–43. DOI: 10.1037/1076-898x.4.1.16 15

[55] Kreifeldt, J. G. and McCarthy, M. (1981). Interruption as a test of the user-computer interface. In *Proc. of Manual Control'19*, pp. 655–667. 12, 13, 15

[56] Kushlev, K., Proulx, J., and Dunn, E. W. (2016). Silence your phones: Smartphone notifications increase inattention and hyperactivity symptoms. In *CHI'16*, pp. 1011–1020, ACM. DOI: 10.1145/2858036.2858359 12, 14

[57] Lane, N. D., Choudhury, T., Campbell, A., Mohammod, M., Lin, M., Yang, X., Doryab, A., Lu, H., Ali, S., and Berke, E. (2011). BeWell: A smartphone application to monitor, model and promote wellbeing. In *Pervasive Health'11*, pp. 23–26, ACM. DOI: 10.4108/icst.pervasivehealth.2011.246161 1, 33

[58] Lane, N. D., Miluzzo, E., Lu, H., Peebles, D., Choudhury, T., and Campbell, A. T. (2010). A survey of mobile phone sensing. *IEEE Communications Magazine*, 48(9):140–150. DOI: 10.1109/mcom.2010.5560598 33

[59] Lathia, N., Pejovic, V., Rachuri, K. K., Mascolo, C., Musolesi, M., and Rentfrow, P. J. (2013). Smartphones for large-scale behaviour change intervention. *IEEE Pervasive Computing*, 12(12):66–73. DOI: 10.1109/mprv.2013.56 47

[60] Latorella, K. A. (1998). Effects of modality on interrupted flight deck performance: Implications for data link. In *HFES'98*, pp. 87–91, SAGE. DOI: 10.1177/154193129804200120 45

[61] Latorella, K. A. (1999). Investigating interruptions: Implications for flightdeck performance. *Technical Report TM-1999-209707*, NASA. 9

[62] Lee, H., Choi, Y. S., Lee, S., and Park, I. (2012). Towards unobtrusive emotion recognition for affective social communication. In *CCNC'12*, pp. 260–264, ACM. DOI: 10.1109/ccnc.2012.6181098 33

[63] LiKamWa, R., Liu, Y., Lane, N. D., and Zhong, L. (2013). Moodscope: Building a mood sensor from smartphone usage patterns. In *MobiSys'13*, pp. 389–402, ACM. DOI: 10.1145/2462456.2483967 33

[64] Lin, J., Mohammed, S., Sequiera, R., and Tan, L. (2018). Update delivery mechanisms for prospective information needs: An analysis of attention in mobile users. In *SIGIR'18*, ACM. DOI: 10.1145/3209978.3210018 37

[65] Lopez-Tovar, H., Charalambous, A., and Dowell, J. (2015). Managing smartphone interruptions through adaptive modes and modulation of notifications. In *IUI'15*, pp. 996–999, ACM. DOI: 10.1145/2678025.2701390 30, 41

[66] Lu, H., Mashfiqui Rabbi, G. T. C., Frauendorfer, D., Mast, M. S., Campbell, A. T., Gatica-Perez, D., and Choudhury, T. (2012). StressSense: Detecting stress in unconstrained acoustic environments using smartphones. In *UbiComp'12*, pp. 351–360, ACM. DOI: 10.1145/2370216.2370270 33

[67] McCrickard, D. S., Chewar, C. M., Somervell, J. P., and Ndiwalana, A. (2003). A model for notification systems evaluation assessing user goals for multitasking activity. *ACM Transactions on Computer-Human Interaction*, 10(4):312–338. DOI: 10.1145/966930.966933 5

[68] McFarlane, D. C. (1997). Interruption of people in human-computer interaction: A general unifying definition of human interruption and taxonomy. Technical report, DTIC Document. DOI: 10.21236/ada333587 8

[69] McFarlane, D. C. and Latorella, K. A. (2002). The scope and importance of human interruption in human-computer interaction design. *Human-Computer Interaction*, 17(1):1–61. DOI: 10.1207/s15327051hci1701_1 5, 9

[70] Mehrotra, A., Hendley, R., and Musolesi, M. (2016a). PrefMiner: Mining user's preferences for intelligent mobile notification management. In *Proc. of UbiComp'16*, pp. 1223–1234, Heidelberg, Germany, ACM. DOI: 10.1145/2971648.2971747 30, 36, 37, 41, 42, 45

[71] Mehrotra, A., Hendley, R., and Musolesi, M. (2016b). Towards multi-modal anticipatory monitoring of depressive states through the analysis of human-smartphone. In *Adjunct UbiComp'16*, pp. 1132–1138, Heidelberg, Germany, ACM. DOI: 10.1145/2968219.2968299 33

[72] Mehrotra, A., Hendley, R., and Musolesi, M. (2017a). Interpretable machine learning for mobile notification management: An overview of prefminer. *GetMobile: Mobile Computing and Communications*, 21(2):35–38. DOI: 10.1145/3131214.3131225 36, 42

[73] Mehrotra, A., Hendley, R., and Musolesi, M. (2019). Notifymehere: Intelligent notification delivery in multi-device environments. In *CHIIR'19*, ACM. DOI: 10.1145/3295750.3298932 38, 39, 43, 45

[74] Mehrotra, A., Muller, S., Harari, G., Gosling, S., Mascolo, C., Musolesi, M., and Rentfrow, P. J. (2017b). Understanding the role of places and activities on mobile phone interaction and usage patterns. *IMWUT*, 1(3). DOI: 10.1145/3131901 31, 46

[75] Mehrotra, A., Musolesi, M., Hendley, R., and Pejovic, V. (2015a). Designing content-driven intelligent notification mechanisms for mobile applications. In *UbiComp'15*, pp. 813–824, ACM. DOI: 10.1145/2750858.2807544 2, 29, 30, 35, 37, 41, 42, 45

[76] Mehrotra, A., Pejovic, V., and Musolesi, M. (2014). SenSocial: A middleware for integrating online social networks and mobile sensing data streams. In *Middleware'14*, pp. 205–216, ACM. DOI: 10.1145/2663165.2663331 33

[77] Mehrotra, A., Pejovic, V., Vermeulen, J., Hendley, R., and Musolesi, M. (2016c). My phone and me: Understanding people's receptivity to mobile notifications. In *CHI'16*, pp. 1021–1032, ACM. DOI: 10.1145/2858036.2858566 2, 3, 31, 35, 36, 41, 46

[78] Mehrotra, A., Tsapeli, F., Hendley, R., and Musolesi, M. (2017c). Mytraces: Investigating correlation and causation between users' emotional states and mobile phone interaction. *IMWUT*, 1(3). DOI: 10.1145/3130948 31, 41, 46

[79] Mehrotra, A., Vermeulen, J., Pejovic, V., and Musolesi, M. (2015b). Ask, but don't interrupt: The case for interruptibility-aware mobile experience sampling. In *UbiComp'15 Adjunct*, pp. 723–732, ACM. DOI: 10.1145/2800835.2804397 46

[80] Miyata, Y. and Norman, D. A. (1986). Psychological issues in support of multiple activities. *User Centered System Design: New Perspectives on Human-Computer Interaction*, pp. 265–284. 2, 5, 21

[81] Monk, C. A., Boehm-Davis, D. A., and Trafton, J. G. (2002). The attentional costs of interrupting task performance at various stages. In *HFES'02*, pp. 1824–1828, SAGE. DOI: 10.1177/154193120204602210 2, 45

[82] Nakayama, M., Takahashi, K., and Shimizu, Y. (2002). The act of task difficulty and eye-movement frequency for the "oculo-motor indices." In *ETRA'02*, pp. 37–42, ACM. DOI: 10.1145/507079.507080 22, 26

[83] Okoshi, T., Nozaki, H., Nakazawa, J., Tokuda, H., Ramos, J., and Dey, A. K. (2016). Towards attention-aware adaptive notification on smart phones. *Pervasive and Mobile Computing*, 26:17–34. DOI: 10.1016/j.pmcj.2015.10.004 33, 42, 45

[84] Pejovic, V., Mehrotra, A., and Musolesi, M. (2015). Investigating the role of task engagement in mobile interruptibility. In *MobileHCI'15 Adjunct*, pp. 1100–1105, Copenhagen, Denmark, ACM. DOI: 10.1145/2786567.2794336 31, 46

[85] Pejovic, V. and Musolesi, M. (2014). Interruptme: Designing intelligent prompting mechanisms for pervasive applications. In *UbiComp'14*, pp. 897–908, ACM. DOI: 10.1145/2632048.2632062 34, 42, 45

[86] Peltonen, E., Lagerspetz, E., Hamberg, J., Mehrotra, A., Musolesi, M., Nurmi, P., and Tarkoma, S. (2018). The hidden image of mobile apps: Geographic, demographic, and cultural factors in mobile usage. In *MobileHCI'18*, ACM. DOI: 10.1145/3229434.3229474 30

[87] Picard, R. W. (1997). *Affective Computing*. MIT Press. DOI: 10.1037/e526112012-054 46

[88] Pielot, M. (2014). Large-scale evaluation of call-availability prediction. In *UbiComp'14*, pp. 933–937, ACM. DOI: 10.1145/2632048.2632060 35, 42

[89] Pielot, M., Cardoso, B., Katevas, K., Serrà, J., Matic, A., and Oliver, N. (2017). Beyond interruptibility: Predicting opportune moments to engage mobile phone users. *IMWUT*, 1(3):91. DOI: 10.1145/3130956 37

[90] Pielot, M., Church, K., and de Oliveira, R. (2014a). An in-situ study of mobile phone notifications. In *MobileHCI'14*, pp. 233–242, ACM. DOI: 10.1145/2628363.2628364 29, 30, 35, 41

[91] Pielot, M., de Oliveira, R., Kwak, H., and Oliver, N. (2014b). Didn't you see my message?: Predicting attentiveness to mobile instant messages. In *CHI'14*, pp. 3319–3328, ACM. DOI: 10.1145/2556288.2556973 19, 34, 42, 45

[92] Rachuri, K. K., Musolesi, M., Mascolo, C., Rentfrow, J., Longworth, C., and Aucinas, A. (2010). EmotionSense: A mobile phones based adaptive platform for experimental social psychology research. In *UbiComp'10*, pp. 281–290, ACM. DOI: 10.1145/1864349.1864393 33

[93] Roegiest, A., Tan, L., and Lin, J. (2017). Online in-situ interleaved evaluation of real-time push notification systems. In *SIGIR'17*, ACM. DOI: 10.1145/3077136.3080808 37

[94] Roegiest, A., Tan, L., Lin, J., and Clarke, C. L. (2016). A platform for streaming push notifications to mobile assessors. In *SIGIR'16*, ACM. DOI: 10.1145/2911451.2911463 37

[95] Rogers, Y. and Marshall, P. (2017). Research in the wild. *Synthesis Lectures on Human-Centered Informatics*, 10(3):i–97. DOI: 10.2200/s00764ed1v01y201703hci037 48

[96] Rosenthal, S., Dey, A. K., and Veloso, M. (2011). Using decision-theoretic experience sampling to build personalized mobile phone interruption models. In *PerCom'11*, pp. 170–187, ACM. DOI: 10.1007/978-3-642-21726-5_11 32, 42

[97] Sacks, H., Schegloff, E. A., and Jefferson, G. (1978). A simplest systematics for the organization of turn-taking for conversation. *Language*, pp. 7–55. DOI: 10.1353/lan.1974.0010 8

[98] Sahami Shirazi, A., Henze, N., Dingler, T., Pielot, M., Weber, D., and Schmidt, A. (2014). Large-scale assessment of mobile notifications. In *CHI'14*, pp. 3055–3064, ACM. DOI: 10.1145/2556288.2557189 2, 3, 30, 36, 41, 42

[99] Smailagic, A., Siewiorek, D. P., Anhalt, J., Gemperle, F., Salber, D., Weber, S., Beck, J., and Jennings, J. (2001). Towards context aware computing: Experiences and lessons. *IEEE Journal on Intelligent Systems*, 16(3):38–46. DOI: 10.1109/5254.940025 40

[100] Speier, C., Valacich, J. S., and Vessey, I. (1999). The influence of task interruption on individual decision making: An information overload perspective. *Decision Sciences*, 30(2):337–360. DOI: 10.1111/j.1540-5915.1999.tb01613.x 8

[101] Tan, L., Roegiest, A., Lin, J., and Clarke, C. L. (2016). An exploration of evaluation metrics for mobile push notifications. In *SIGIR'16*, ACM. DOI: 10.1145/2911451.2914694 37

[102] Wang, T., Cardone, G., Corradi, A., Torresani, L., and Campbell, A. T. (2012). Walk-Safe: A pedestrian safety a page for mobile phone users who walk and talk while crossing roads. In *HotMobile'12*, pp. 5–14, ACM. DOI: 10.1145/2162081.2162089 1

[103] Weber, D., Voit, A., Auda, J., Schneegass, S., and Henze, N. (2018). Snooze!: Investigating the user-defined deferral of mobile notifications. In *MobileHCI'18*, ACM. DOI: 10.1145/3229434.3229436 46

[104] Weber, D., Voit, A., Kratzer, P., and Henze, N. (2016). In-situ investigation of notifications in multi-device environments. In *UbiComp'16*, pp. 1259–1264, ACM. DOI: 10.1145/2971648.2971732 37, 43, 45, 47

[105] Weiner, B. (1965). Need achievement and the resumption of incompleted tasks. *Journal of Personality and Social Psychology*, 1(2):165–168. DOI: 10.1037/h0021642 15

[106] Weiser, M. (1991). The computer for the 21st century. *Scientific American*, 265(3):94–104. DOI: 10.1038/scientificamerican0991-94 14, 15

[107] Weiser, M. and Brown, J. S. (1997). *The Coming Age of Calm Technology*. Springer. DOI: 10.1007/978-1-4612-0685-9_6 15

[108] Westermann, T., Möller, S., and Wechsung, I. (2015). Assessing the relationship between technical affinity, stress and notifications on smartphones. In *MobileHCI'15 Adjunct*, pp. 652–659, ACM. DOI: 10.1145/2786567.2793684 30, 41

[109] Zabelina, D. L., O'Leary, D., Pornpattananangkul, N., Nusslock, R., and Beeman, M. (2015). Creativity and sensory gating indexed by the P50: Selective vs. leaky sensory gating in divergent thinkers and creative achievers. *Neuropsychologia*, 69:77–84. DOI: 10.1016/j.neuropsychologia.2015.01.034 11

[110] Zeignaric, B. (1927). Das behalten erledigter und unerledigter handlungern. *Psychologische Forschung*, 9:1–85. 11, 12

[111] Zijlstra, F. R., Roe, R. A., Leonora, A. B., and Krediet, I. (1999). Temporal factors in mental work: Effects of interrupted activities. *Journal of Occupational and Organizational Psychology*, 72(2):163–185. DOI: 10.1348/096317999166581 2, 12, 14

Authors' Biographies

ABHINAV MEHROTRA

Abhinav Mehrotra is a Machine Learning Engineer at Samsung AI Center, Cambridge, UK. He obtained his Ph.D. in Computer Science from the University of Birmingham, UK, where he worked on intelligent mobile notification systems. He joined University College London (UCL) as a postdoctoral researcher, where his research focused on behavior modeling and digital health through the analysis of contextual information obtained via embedded sensors. After his postdoctoral work at UCL, he joined Samsung AI Center, where his research efforts are toward optimization of machine learning models in order to support on-device AI systems, and design of intelligent speech interaction systems.

MIRCO MUSOLESI

Mirco Musolesi is Full Professor of Data Science at University College London (UCL) and a Turing Fellow at the Alan Turing Institute. He is also Full Professor of Computer Science at the University of Bologna. At UCL he leads the Intelligent Social Systems Lab. He received a Ph.D. in Computer Science from UCL and a Masters in Electronic Engineering from the University of Bologna. After postdoctoral work at Dartmouth College and Cambridge, he held academic posts at St Andrews and Birmingham. Over the past years, the focus of the work of his lab has been the design of next-generation intelligent systems mainly based on computational and mathematical models of human behavior and social dynamics. More recently, he has been interested in designing autonomous systems, possibly with humans in the loop. He is interested in both theoretical and systems-oriented aspects of these research areas. He works at the interface of several disciplines including Ubiquitous Systems, Autonomous Systems, Machine Learning, Artificial Intelligence, and Computational Social Science.